General Editor: Robin Gilmour

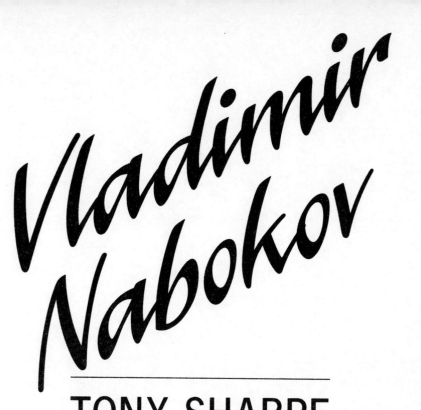

TONY SHARPE

Lancaster University

Edward Arnold
A division of Hodder & Stoughton
LONDON NEW YORK MELBOURNE AUCKLAND

© Tony Sharpe

First published in Great Britain 1991

Distributed in the USA by Routledge, Chapman and Hall, Inc.
29 West 35th Street, New York, NY 10001

British Library Cataloguing in Publication Data

Sharpe, Tony
   Vladimir Nabokov
   I. Title
   823

ISBN 0-7131-6575-8

Typeset in 10/12pt Linotron Sabon by
Hewer Text Composition Services, Edinburgh
Printed and bound in Great Britain for Edward Arnold,
a division of Hodder and Stoughton Limited,
Mill Road, Dunton Green, Sevenoaks, Kent TN13 2YA by
Biddles Ltd, Guildford & King's Lynn

Contents

For Jane and Eva

General Editor's Preface

Fiction constitutes the largest single category of books published each year, and the discussion of fiction is at the heart of the present revolution in literary theory, yet the reader looking for substantial guidance to some of the most interesting prose writers of the twentieth century – especially those who have written in the past 30 or 40 years – is often poorly served. Specialist studies abound, but up-to-date maps of the field are harder to come by. *Modern Fiction* has been designed to supply that lack. It is a new series of authoritative introductory studies of the chief writers and movements in the history of twentieth-century fiction in English. Each volume has been written by an expert in the field and offers a fresh and accessible reading of the writer's work in the light of the best recent scholarship and criticism. Biographical information is provided, consideration of the writer's relationship to the world of their times, and detailed readings of selected texts. The series includes short-story writers as well as novelists, contemporaries as well as the classic moderns and their successors, Commonwealth writers as well as British and American; and there are volumes on themes and groups as well as on individual figures. At a time when twentieth-century fiction is increasingly studied and talked about, *Modern Fiction* provides short, helpful, stimulating introductions designed to encourage fresh thought and further enquiry.

Robin Gilmour

Preface

My initial debt of gratitude is to the friends who introduced me to Nabokov's work, and with whom I shared my first enthusiasm; then to my teachers at undergraduate and graduate level, under whose guidance I developed that interest. I am also grateful to the General Editor, Robin Gilmour, for inviting me to write this volume in the series; I have enjoyed doing so, and hope some of that enjoyment communicates itself to my readers. My chief thanks are due to the book's dedicatees: to Jane, who read it in manuscript and rescued me from some infelicities; and to our daughter Eva Bessie, who grew smilingly as the book was being written, and permitted her father more sleep than a new parent has any right to expect.

A Note on Editions

I have used the Penguin editions for this study, apart from *The Enchanter*, which is published by Picador. Nearly all of Nabokov's important fiction is currently available in Penguin, an exception being *The Defence*, which is published in paperback by Oxford University Press. The principal publisher of Nabokov's work in the UK is Weidenfeld and Nicolson; in the US the principal publishers are Putnam and McGraw-Hill.

If this be magic, let it be an art
Lawful as eating.

The Winter's Tale

It all comes back to that, to my and your 'fun' – if we but allow the term its full extension; to the production of which no humblest question involved, even to that of the shade of a cadence or the position of a comma, is not richly pertinent.

Henry James, Preface to *The Golden Bowl*

1

The Life and its Times

When Vladimir Nabokov died in a Lausanne hospital on 2 July 1977, there seemed a grim appropriateness in the fact that he had been struck down by a nameless virus. For such had been his accomplishment as linguist and as word-magician, that few would have doubted his ability to master anything that could be met with in a dictionary: therefore death's agent, unidentified, had had no word by which to be pinned down and, as a non-verbal phenomenon, negated its victim's formidable powers. This was not to be the only Nabokovian aspect of the 78-year-old writer's end. For he left uncompleted his *Original of Laura,* which his son asserted 'would have been Father's most brilliant novel'; and the straining conditional-perfect tense of this declaration, its flight toward the future impeded by the gravitational pull of time, also seems appropriate in the case of Nabokov: whose fictions constantly evoke realms of possible completion, hovering beyond the inevitability of final paragraphs.

Nabokov's ghost might well take exception to his inclusion in a series such as this, in the company of writers whom he had not personally selected, and whose implied equality with himself he would contemptuously repudiate. During his lifetime he refused to share plat- forms with literary and other 'celebrities' whose merit he doubted, and indeed was averse to presenting himself as a public figure through any other medium than that of his writing (and, latterly, through interviews which he had scripted). Nabokov's attitude may be glimpsed behind the retort made by his character Adam Krug, eminent philosopher, on hearing himself described as one of the very few celebrities in a small country: 'who are the other stars of this mysterious constellation?'. So superb a posture was easily taken for arrogance, and led to the exasperated picture painted by Edmund Wilson, as the friendship between these two men of letters foundered: in which he suggested that Nabokov viewed himself as 'unique and incomparable', and that anyone disputing his judgement was likely to be dismissed as 'an oaf and an ignoramus, . . . usually with the implication that he is also a low-class person and a ridiculous personality'.[1]

Why was he proud – if proud he was? By the time he died Nabokov had composed seventeen novels in both Russian and English, and several collections of short stories and poems, as well as an immensely scholarly and equally controversial translation (with two-volume commentary) of Pushkin's verse-novel masterpiece *Eugene Onegin*, together with an impressively original volume of auto-biography, entitled (finally) *Speak, Memory*. After initial notoriety his novel *Lolita* (1955) had brought him wealth and fame, and his works were translated, with as much personal supervision as possible, into many languages. If he was a celebrity at his end, in his beginning also Nabokov had been distinguished: born the eldest son to a noble Russian family in St Petersburg, before his eighteenth birthday he had inherited from an uncle an estate valued at two million pounds. Before his nineteenth birthday this had all been swept away by the Bolshevik revolution; and before his twentieth birthday Nabokov had left the country of his birth forever, to embark on the uncertain and for the most part penurious existence of an émigré in Germany, France and the USA, where he acquired citizenship and, eventually, the best-selling success that enabled him to return to Europe at the age of sixty to live, in a more opulent quasi-statelessness, in an hotel at Montreux, Switzerland.

The effect of any writer's formative years is likely to be incalculably large; and in Nabokov's case, his birth and growth to early manhood in that setting of pre-Soviet Russia, which contrasted so sharply with the years that followed for him, are a constant element in his fiction. From a childhood and youth that had unfolded in the spacious plenty of large rooms in big houses surrounded by rambling parks, its rhythms those of the seasonal transition between town house and country estate, punctuated by trips abroad in Pullman carriages, Nabokov was translated to a manhood of frequently-changed furnished lodgings and chronically modest means. Born into a respected and influential family, the favourite child of loving parents, inheritor of a dynastic

---

1. These words come from Wilson's review of Nabokov's translation of *Eugene Onegin* in the *New York Times Book Review*, 15 July 1965. The correspondence between the two men during the period 1940–71 has been published as *The Nabokov – Wilson Letters* (London: Weidenfeld & Nicolson, 1979), edited and introduced by Simon Karlinsky; this volume will hereafter be cited in the running text thus: (NWL p. –).

perspective looking back through centuries and a future prospect of equally splendid stability, Nabokov would see that family scattered, itinerant, displaced: his father assassinated at a political meeting in Berlin, his mother dying in Prague, a sad widow whose reduced circumstances his own feeble finances had not much alleviated. He was not the man to forget such losses, and in various ways the note of severance is sounded in Nabokov's writing; but he was not animated by a remembrance of roubles past. The loss of riches did not affect him, so much as the loss of people, places, associations; textures of time and life whose sacrifice attested to a world of waste. To counter this waste, his fiction is an act of imaginative recuperation, lit by a spirit not vengefully disappointed but comically, quizzically affirmative.

At the turn of the century St Petersburg was a city with a rich cultural life. Proudly outward-looking, its inhabitants took a condescending view of Moscow (a city which Nabokov never visited) as rather provincial, by contrast with their own sense of being part of Europe – facilitated by the express train which left for Paris twice each week. Nabokov was the eldest of five surviving children (two brothers, two sisters); their father, V. D. Nabokov, was a respected statesman and active member of the liberal aristocracy, whose reformist tendencies led him to fall foul of the ill-fated Tsar Nicholas II: he was briefly imprisoned in 1908. The household over which he presided was a model of well-educated cosmopolitanism, in which French and English were spoken as naturally as Russian – indeed, more so: by his own account Nabokov learnt to read English before he could read Russian. He subsequently remarked to the effect that his was a perfectly ordinary tri-lingual childhood; and if he accepted it as normal that his family owned many thousands of acres, it was also normal that his father owned a library of some 10,000 volumes, necessitating the employment of a family librarian. Looking back from 1972, Nabokov supposed himself to have had 'the happiest childhood imaginable'.

This had taken place, however, against an increasingly agitated backcloth of contemporary events – in which his father took an active part – as desire for political reform in Russia encountered an inflexible authoritarianism at the centre. The Bolshevik coup in late 1917 dashed liberal hopes of an enlightened reformed constitution, and the Nabokovs fled St Petersburg for the Crimea, where they passed more than a year before sailing out of Sebastopol harbour in April 1919, the shabby Greek freighter on which they escaped being strafed by Bolshevist machine-guns. Supported by a few jewels they had managed to snatch up on departure, the family spent a short while in London before moving to Berlin, where the cost of living was lower and there was a large expatriate Russian community. Nabokov and his slightly younger brother Sergey attended Cambridge University together; and in March 1922, the year of his graduation, Nabokov had been reading poetry aloud to his mother in their Berlin apartment, when news came that at a meeting elsewhere in the city V. D. Nabokov had been shot dead by the accomplice of a drunken pro-Tsarist assassin, whom he was preventing from making good an attempt on the life of the guest speaker (a former government minister in Russia).

His mother and her younger children moved to Prague for financial reasons, but Nabokov stayed in Berlin, where he met and married his wife Véra, and began to make a name (or more accurately, a pseudonym, 'Sirin') for himself as a writer, amongst the expatriate community. To make ends meet he gave language lessons and tennis-coaching to well-to-do Berlin families, but money was always tight. As Nazism gained ground the wisdom of remaining in Germany became ever more questionable, and in 1937 the Nabokovs and their only child, Dmitri, managed to remove themselves to France, settling eventually in Paris, by now the focus of Russian expatriate life. These were difficult years, in which the problem of making a living was enlarged by Nabokov's growing realization that he would need to change the language in which he wrote; for although initially the displaced Russians abroad could constitute a sizeable cultural community, as time passed, taking with it the hopes of any return to a non-Bolshevist Russia, the coherence of this group diminished. With the virtual sealing-off of the Soviet Union from cultural congress with the West, émigré writers such as 'Sirin' saw that they were doomed to semi-extinction, as their only available audience dispersed; and this, as Nabokov later put it, meant that 'the whole thing acquired a certain air of fragile unreality'.[2] Already he was looking toward an English audience: his novel *Camera Obscura* (later *Laughter in the Dark*) had been translated, and he himself had made an English translation of *Despair* before leaving Germany (published in England in 1937). The novel on which he was working, *The Gift*, was to be his last (and many think his best) published in Russian; he completed it at the beginning of 1938, and at the end of that year began work on the first of his novels in English, *The Real Life of Sebastian Knight*.

2. *Speak, Memory: An Autobiography Revisited* (Harmondsworth: Penguin, 1969), p. 215. This is the revised version of Nabokov's autobiography – first published in New York (1951) as *Conclusive Evidence* – hereafter cited in the running text thus: (SM p. 215).

He was looking with increasing urgency for an academic position, at first in England, but when none came he turned towards America. In this he was at last successful, obtaining through a contact a short-term teaching job at Stanford University, which carried with it the all-important visa that would enable him to leave France. There were further bureaucratic and financial obstacles to be overcome, but the family finally set sail from France in May 1940. Hitler's armies had already invaded the Low Countries and were speeding towards Paris; in a repeat of his narrow escape from Sebastopol, Nabokov's liner was one of the last to leave before the ports were closed by the occupying Nazis. In America he embarked upon a series of short-term academic appointments, utilizing his skills as a lepidopterist as well as his skills as linguist and man of letters, eventually coming to rest at Wellesley and then, more securely, at Cornell University, where he rose to the rank of full professor. During this time he and his family lived in a succession of temporary homes, usually the houses of colleagues on leave, which appear in various guises in his fiction of this period. By the time he had become a full professor it would have been possible for Nabokov to think of buying his own house; and when the success of *Lolita* made him once more wealthy, he could have become a substantial man of property; but he never owned a house outside Russia. He resigned from Cornell, leaving the USA toward the end of 1959, and settled with his wife at the Montreux Palace hotel, where he lived until his death. Life in the hotel may have simulated some of the conditions of his far-off Russian days; it may have expressed his disinclination ever to trust the things of this world to the extent of home-ownership; and it may also have helped to conceal from Nabokov himself the fact that, despite his declared inclination to do so, he would not return to live in America.

An 'interview' he gave in 1962 contains a typical pronouncement, in reply to a question asking for his comment on an aspect of the contemporary novel: 'I am not interested in groups, movements, schools of writing and so forth. I am interested only in the individual artist'. This is a constant feature of Nabokov's response to literature, and of his response to critics who attempted to set his own work in this or that literary tradition. It can well be seen that at the most basic level a life and a career such as his present difficulties of classification: nine of his novels were originally composed in Russian, eight in English; should we therefore regard him as a Russian or an English-American writer? Decontextualised by the accidents of history, Nabokov did what he could further to decontextualise himself, indignantly disclaiming various literary kinships foisted upon him by incautious commentators, insisting rather on his own uniqueness. With regard to those same accidents of history that had so radically affected the story of his life, it may seem that he displays a sensibility almost pathologically dissociated: in October 1917 in St Petersburg, the budding poet finished producing his evening quota of lines, and noted that during their composition 'fierce rifle fire and the foul crackle of a machine gun' had been audible from the street outside;[3] in mid-May 1940, as Hitler's panzer regiments stormed into the Low Countries, Nabokov in Paris brought to loving completion his very best, most demonically teasing, chess problem (SM pp. 224–5).

These and similar episodes are not, of course, without a certain self-consciousness on Nabokov's part; but even so there is a gulf between his father's profound commitment to affairs of state, and his own apparent indifference to most manifestations of the public world – an indifference that is all the more remarkable when we consider the effects that world had had upon his personal circumstances. Nabokov's aloofness from current events and his absorption in the problems of fictional composition ally him with some of the Modernist writers – most particularly Joyce – for whom the work of art was to a high degree an independent world. Yet the conditions in which he composed his earlier novels (in particular) were beset by the contingent world: for him no Jamesian summer-house, no Proustian cork-lined room, but a bathroom doubling up as study. A more significant difference is that although by his expatriation Nabokov partook of the statelessness common amongst Modernist writers, he actually had a very strong sense of home, which survived his exile. Spiritually, he did not lose his roots; it was, rather, the tenacity with which he preserved them in alien soils that led to his apartness. Notwithstanding the Anglophilic nature of his upbringing, Nabokov derived few benefits from Cambridge, where he did not thrive; looking back, he saw that the story of his years there was 'really the story of my trying to become a Russian writer'; and in Germany too, he felt the need to protect the fragility of his transplanted Russianness from being impaired by its surroundings. This explains why, in spite of the fact that he spent ten years in Berlin before Hitler took over in 1933, his evocations of that city and that period seem so monochrome, by contrast with the gaudy memories of other writers, there in voluntary exile: Nabokov transmits no sense that it was an exciting place to be in. Really, nowhere that was not Russia could have done; and in his places of exile, he was further afflicted by the widespread failure to understand what had

3. Quoted by Brian Boyd in *Vladimir Nabokov: The Russian Years* (London: Chatto & Windus, 1990), p. 133. Hereafter cited as Boyd.

actually happened to his homeland; for many in the West, Lenin and even Stalin looked like exciting visionaries rather than harbingers of terror and disappropriation.

For Nabokov, what counted in an individual life were the inner rather than the exterior factors, internal conditions rather than outside conditioning. He insisted that his opposition to the Soviet system was not rooted in any regret for lost property: 'the nostalgia I have been cherishing all these years is a hypertrophied sense of lost childhood, not sorrow for lost banknotes' (SM p. 59). He went so far as to hint his gratitude to the Russian revolution, for the lesson it enforced about the vulnerability of 'real estate', implying to him that the unreal estate of art was a more reliable domain. His enforced exile may also have given depth and specificity to a vision initially inclined to content itself with languorous generalities; as his biographer Brian Boyd makes clear, the poetry young Vladimir composed was heavy with a sense of loss and sombre memory, well before his actual dispossession took place. Like many youthful writers (not all of whom necessarily grow up), Nabokov had been in search of feelings appropriate to his vocabulary; events were to lend a rough hand to his seeking. But if the Muse of History in this way helped flesh out his sense of loss, her intrusion would have gone for nothing without the operation in him of her sisters Memory and Imagination (two Cinderellas to one uglier sibling): life takes a shape passively from history, but actively shapes itself through memory and imagination.

This much is clear from *Speak, Memory*, Nabokov's autobiography dealing with the years up to his emigration to America (a subsequent volume was projected but not completed). It is a remarkable piece of writing, shot through with the interactions of happenstance with design, and is probably one of his most approachable works – perhaps because generically it seems least deviant: the portrait of his Swiss governess, for example, has the reassuring feel of an old-fashioned character study (and was one of the first chapters to be written). Yet, considered as a memoir or autobiography, the book is as remarkable for what it omits, as for what it reveals: tutors, governesses, teachers, and chance acquaintances are described more vividly than more immediate members of the family, who are evoked as emotional quantities rather than given as portraits. Albeit that the story is about Nabokov, he remains in partial shadow; he seems to have followed Hawthorne's advice, given at the beginning of *The Scarlet Letter,* about how a reticent author should proceed in such matters: by imagining that he addresses a close friend (*Speak, Memory* is actually addressed to Véra Nabokov) the situation admits a certain confidentiality of tone whilst equally permitting the author to 'keep the inmost Me behind its veil'. Such retention of the 'inmost Me' owed more to Nabokov's sense of decorum than to a fearful masking of the subconscious and its primal scenes.

Indeed, the unconstructed self was of little interest to him. By contrast with Proust's monumental *In Search of Lost Time* (the first volume of which Nabokov nominated along with *Ulysses*, Kafka's *Metamorphosis*, and Bely's *Petersburg* as one of the pre-eminent masterpieces of twentieth-century fiction), *Speak, Memory* does not present memory at its deepest as an involuntary impulse, associations triggered off by chance sensation. It is, rather, a conscious remembering that the book celebrates and enacts, in which even the sudden recalling of the name of his first girlfriend's dog is, despite the simulacrum of spontaneity, produced by an effort of recollection (SM p. 119): in Nabokov, memory is much more like a retrieval system than it is like a bran-tub. *Memino, ergo sum* might be the motto of *Speak, Memory*; and it should be borne in mind, especially when considering the book's triumphant artfulness, that the validity of the recollecting is not to be measured simply in terms of its factual accuracy. The further we read, the more it is borne in upon us that 'reality' is consequent upon a mode of perception rather than existing as an inert assemblage of facts, and that what we have before us is a mind seen in the act of fashioning a life.

The first part of the first chapter recognizes the prison of time into which the new soul is born, the inter-implication of birth and death; but it also celebrates the dawning of consciousness as self-consciousness, and the more important birth recorded here is not the moment of mammalian parturition but the birth of young Vladimir's sense of self, aged four. When, much later, he was asked what distinguishes man from animals, his answer was, 'being aware of being aware of being. In other words, if I not only know that I *am* but also know that I know it, then I belong to the human species. All the rest follows – the glory of thought, poetry, a vision of the universe'.[4] In his memoir much emphasis is placed on the glory of our mental life and its ardent constructiveness: 'How small the cosmos (a kangaroo's pouch would hold it), how paltry and puny in comparison to human consciousness, to a single individual recollection, and its expression in words!' (SM p. 21). Inevitably, the consideration of such contrast also suggested a dark conclusion, as when, near the end of the book, Nabokov refers to 'the utter degradation, ridicule, and horror of having developed an infinity of sensation and thought within a finite existence' (SM p. 227); but equally, he asserts that

---

4. *Strong Opinions* (London: Weidenfeld & Nicolson, 1974), p. 142. Hereafter cited in the running text thus: (SO p. 142).

it is only through sensation and thought that human finitude can be surpassed: 'It is certainly not then – not in dreams – but when one is wide awake, at moments of robust joy and achievement, on the highest terrace of consciousness, that mortality has a chance to peer beyond its own limits' (SM p. 41).

Speak, Memory, like the Bible, starts with a version of paradise; but unlike the religious account, expulsion from this Eden is not consequent upon any 'fall' into consciousness, still less any 'fall' into sexuality, both of which had been triumphantly accommodated: instead it follows the fall into politics, the obtrusive imposition of a wholly regimented mode of being onto an entire nation. The 'lost childhood' fuelling Nabokov's nostalgia in exile was not simply his own, but in some degree suggests the loss of innocence, the failure of expansive curiosity, institutionalized by the Soviet system: whose citizens and children alike were henceforth to be shut up in a great grim schoolroom, with the savagest of punishments and the briefest of holidays. The particular nature of Nabokov's exile was that it comprised both temporal and geographical disjunction: in Tsarist times, writers exiled to Siberia could dream of returning to a recognizable land; whereas Nabokov, even had he managed to revisit the scenes of his happy childhood and youth, would have found it all changed utterly, and a terrible ugliness prevailing. Not only miles, but years, would have to be retraced; and this conscious recapturing of the past, the attempted reinhabitation of his lost domain in its most important, immaterial aspect, is what *Speak, Memory* undertakes.

His was no soft-focus imprecision of nostalgia, but an intensive, intensively specific passion, which hungered for and celebrated details. In a much-quoted passage from *Rasselas* (1759), Dr Johnson defined 'the business of the poet' as concerning 'not the individual, but the species; to remark general properties and large appearances: he does not number the streaks of the tulip, or describe the different shades in the verdure of the forest'. This Augustan aesthetic is almost the exact obverse of Nabokov's own theory and practice; consciousness was not for him an abstract generality, but predicated a single self in a particular relation with an objective world: *not* to discriminate the forest's various greennesses would amount to a betrayal of that self, that world. Another betrayal, so far as he was concerned, recurred on a nightly basis: when the mind, slipping into neutral gear, disengages from externals and surrenders to the fatuity of dreams. 'Sleep is the most moronic fraternity in the world, with the heaviest dues and the crudest rituals,' he declared; 'no matter how great my weariness, the wrench of parting with consciousness is unspeakably repulsive to me' (SM p. 85). He was himself a virtual insomniac; sleep, as the effacement of individuality, destroyed the distinction between men that only consciousness sustains and – at the opposite remove from our awareness of being aware of being – is something that we share with animals.

In the light of all this, Nabokov's disgust for Communist politics and his refusal to believe in 'schools of writing and so forth' are seen to follow logically from his profound aversion to collectivism. Similarly, his passionate belief in the importance of minute particulars, in the validity of numbering the streaks of tulips, is evidenced in his pursuit of lepidoptery. Chapter Six of *Speak, Memory* is devoted to his interest in butterflies, which began when he was seven, and which he pursued to the extent of publishing papers in scientific journals; its most intensive period fell during the 1940s, when he was Research Fellow in Entomology at Harvard's Museum of Comparative Zoology. There, his work was mostly devoted 'to the classification of certain small blue butterflies on the basis of their male genitalic structure'; this involved many hours of microscopic scrutiny, using character-istics of the subject quite invisible to the naked eye to discriminate between specimens. Working at the frontiers of knowledge in this way, Nabokov knew the excitement of original discovery, the ecstasy of perception attendant on his observation of some feature hitherto known only to a hypothetical Creator. In Keats's sonnet 'On First Looking into Chapman's Homer', the poet (himself a former medical student) has brilliantly paralleled artistic with scientific discovery: 'Then felt I like some watcher of the skies / When a new planet swims into his ken'; Nabokov urged his students in America to combine the artistic and scientific temperaments, in order to become good readers. In 1962 he defined his 'pleasures', half-humorously, as 'the most intense known to man: writing and butterfly-hunting'.

We might suppose a preoccupation with the taxonomy of butterflies to be the mark of a mind pedantically concerned with inert forms of tabulation, keen to encase nature's proliferation in strait-jacket categories. There is also a potentially repellent difference between the living butterfly we glimpse and the trayful of specimens, impaled, englassed, and dead. Nabokov's own account of the connections between art and science, however, emphasizes that his research in lepidoptery revealed to him the creative excesses of nature: how the patterning on a butterfly's wing, for example, in the accuracy of its mimicry fantastically exceeded the visual competence of the predator it was supposed to discourage (refuting any theory of 'natural selection'). 'I discovered in nature,' he wrote, 'the nonutilitarian delights that I sought in art. Both were a form of magic, both were a game of intricate enchantment and deception' (SM p. 98); it was nature's apparent delight in its own forms, beyond any discoverable practical purpose, that he stressed, and which in some ways is equivalent to the prodigal superabundance of human consciousness – whose lack of direct application bespeaks its profoundest value and freedom.

Nabokov's interest in chess obviously conforms to his pleasure in playing intricate games; and it can readily be seen how an attachment to the 'nonutilitarian' aspects of human consciousness would set him at odds with Soviet Russia's (or anybody else's) Marxist ideology, whereunder thinking not directly applicable to social reality was dismissed as 'false consciousness'. It is small wonder that he greatly admired *Hamlet,* that undisciplined play whose hero cannot bear to leave off the fantastications of his agile mind; and in contrast to the predictable plots of authorized Soviet fiction – the heroic tractor driver, the selfless factory-worker, the fearless exposer of spy or saboteur – Nabokov's best novels teem with digressive energy. One of the many great moments in *King Lear* occurs when Lear protests against a view of life based on calculations of necessity; 'O reason not the need!' he cries, 'Allow not nature more than nature needs, / Man's life is cheap as beast's' (II. iv). In less tragic mode, Nabokov would agree that the noble aspect of human life derives from what it adds to basic biological requirements; and that this capacity to respond to impulses unconnected with imperatives of simple survival is, unfunctionally creative, an absolutely essential element of our humanity. Paradoxically, by being useless it is indispensable, and art, as the plaything of superfluous consciousness, has a necessary unnecessariness.

The verbal inventiveness and extravagance that characterizes Nabokov's writing is part of this excess, as is its tendency to shift between languages. But there was a darker, potentially desperate side, as he was well aware: for his exile from Russia became, implacably, an estrangement from Russian. He was at first insecure enough in his English to seek help in matters of grammar and usage from native speakers; and latterly, when his own fame and a relative thaw in East-West relations made possible the circulation of some of his works inside the Soviet Union, his own translation into Russian of *Lolita* was felt by some to be a disappointing and inferior performance. His eventual stylistic brilliance in English was perhaps only achievable by one who *used* the language rather than inhabited it, and who brought an outsider's ear and eye to its potentialities. His English style has been criticized for its uneasy combination of English Literariness with American demotic, and he was perhaps too ready to rely upon a dictionary; but it is surely the case that, in addition to a job and a national predisposition to be hospitable, America offered

Nabokov a language whose fertile informalities were being shaped by the modifications of its many citizens, like himself, whose mother tongue was not English.

In this as in other aspects of Nabokov's life we see the interweaving of accident and pattern. Near the beginning of *Speak, Memory* he declares that 'the following of such thematic designs through one's life should be, I think, the true purpose of autobiography'. The book abounds with artful conjunctions and novelistic foreshadowings of later events – a particular example is the various intimations of his father's untimely death (and since the chapter devoted to his father ends with his *not* having been killed in a duel, as his son had been fearing, this in turn forms part of the theme of false death running through Nabokov's fiction). Beginning with its scene of birth that simultaneously evokes the cradle and the coffin, *Speak, Memory* ends with the rebirth of the Nabokovs as, new child in hand in their final European garden, they approach the ship that will carry them to transatlantic safety. In a different key, the young Vladimir's aspirations to being a great Russian writer are wittily and unobtrusively signalled, by synchronizing the onset of his potency with a situation falling vacant on Parnassus: he asks his father about erections as his father reads in the newspaper of the death of Tolstoy (SM p. 162). But he was not to be to Russia what Tolstoy was; and when in 1971 he was asked to define himself, he replied that he saw himself 'as an American writer raised in Russia, educated in England, imbued with the culture of Western Europe' (SO p. 192).

Treating in turn the elements he proposes here, we can see that a literary tradition as formally elastic as America's could happily accommodate his work: a canon that found room for *Moby-Dick* and *The Scarlet Letter* need not strain at *Pale Fire*; all three have in common fictional self-consciousness and a high degree of experimentalism. Nabokov did not, however, share the culturally embattled situation of a Melville or a Hawthorne, Modernist writers before their time. In his continuous concern with the art of fiction, and in his musings on the relations between that art and 'life', he has some resemblances to Henry James (not least in the prefaces he composed for his translated novels, drawing the reader's attention to supposed felicities); but he had little time for James as a practitioner – he preferred the work of Edgar Allan Poe. For all that, he refused to consider Poe's distant heir, William Faulkner (who could write English as if it were a foreign language, in spite of having no other), as a 'major' novelist. Amongst his American contemporaries he singled out for mention writers not obviously like himself, Salinger and Updike; and if we were to nominate writers who have a degree of affinity with him (not the same thing as showing his influence), we might think that Thomas Pynchon's metafictional preoccupations ally him with Nabokov (who taught Pynchon at Cornell), and we might also consider a writer like Robert Coover, who like Nabokov has a strong sense of the essential irresponsibility of fiction, as well as of its fixations (for examples, *Spanking the Maid* or *Gerald's Party*). Whatever the appropriateness of such comparisons, the strength of Nabokov's Russian colouring and the importance of that past to him make it difficult to assent to the proposition that he is primarily an American writer.

With the gradual rectification of our Western ignorance about Russian cultural and political history in the early years of this century, it is becoming clearer how important to Nabokov was the literary climate of St Petersburg, in which he grew up. Simon Karlinsky asserts that 'the spectacular explosion of Russian literary creativity after 1905' (and before the Bolsheviks shut up the shop and falsified its accounts) was an essential background to Nabokov's art, whose origins he traces to 'the experimental prose of Remizov and Bely, . . . the more traditionalist, but stylistically exquisite prose of Bunin and . . . the great and innovative poetry that was then being written by Annensky, Blok, Bely and, later, Mandelstam and Pasternak' (NWL p. 20). From an earlier period of Russian literature, it is clear that some of the features Nabokov openly admired in a writer like Gogol or, historically later, Chekhov (another artist/scientist), can be glimpsed in his own work: but 'influence' seems too coarse a concept here. Of the pre-eminent nineteenth-century Russian novelists, Nabokov held Tolstoy in the highest regard and Dostoevsky in the lowest contempt; to no avail, commentators pointed out that he, like Dostoevsky, made use of the theme of the 'double'. Nabokov's dismissal of Dostoevsky is interesting, in the light of a lecture on modern Russian prose given in 1984 by the Russian poet Joseph Brodsky (recently expatriated and subsequently a Nobel laureate); for Brodsky contended that the impoverishment of contemporary writing had been due to its failure to capitalize on its legacy from Dostoevsky, and its concentration on the legacy from Tolstoy: 'the road not taken was the road that led to modernism, as is evidenced by the influence of Dostoevsky on every major writer this century, from Kafka on. The road taken led to the literature of socialist realism'.[5]

Whether or not he is right about Dostoevsky, it is significant how little attention Brodsky devotes to Nabokov; for it suggests that, whatever the links with earlier Russian writers (as outlined

5. See Brodsky's essay 'Catastrophes in the Air' in his *Less Than One: Selected Essays* (Harmondsworth: Penguin, 1987), p. 280.

by Karlinsky, and suggesting an alternative antecedence for Russian Modernism), Nabokov is not perceived as centrally relevant to later Russian writing. Nadezhda Mandelstam, the poet's widow, made the following comment on the Russian *Speak, Memory*:

> Oddly enough – and though I do not like him – I can forgive only Nabokov's somnambulist excursion into his childhood. Separated from his native country, no longer immersed in its language and history – and having lost his father in the way he did – Nabokov re-creates the idyll of his childhood as his only link with the country of his forebears. Living the life of an expatriate, he was deprived of the chance of coming to maturity.[6]

These are quite hard words, implying that Nabokov had been fundamentally impaired as a writer by his expatriation. Whether or not she is correct, what she does bring out is the dreadful psychic wounding that expatriation could entail, and this explains how it was that so many who had escaped the Red advance returned voluntarily, to a variety of fates (the poet Anna Akhmatova, the novelist Bely, the formalist critic Shklovsky and the composer Prokofiev come to mind). Unwavering as was his detestation of the Soviet system and its consequences, Nabokov's absence from the scenes of Russia's suffering, particularly under Stalin but also under previous and subsequent dictators, meant that he could not – like Pasternak and then Solzhenitsyn – function as the voice or conscience of the nation. In the current climate of *glasnost* his work is becoming widely available behind the rusty remains of the Iron Curtain; but although Russian readers may come retrospectively to regard Nabokov as the literary equivalent of a government in exile, who kept artistic faith during the Communist nightmare and its cultural impoverishment, Nadezhda Mandelstam's opinion (if representative) gives cause for doubt.

Although published in England before America, until *Lolita* Nabokov made little impression here. With regard to English literary history, his position is made unusual in that the writers he was familiar with in his youth, and whom he seems to have admired most, were somewhat old-fashioned practitioners such as Rupert Brooke, H. G. Wells, and Robert Louis Stevenson. It is doubtless another proof of the fatuity of conventional literary groupings, that one who was himself a considerable innovator should have been so indifferent to its modernist exponents in English (with the notable exception of

6. Nadezhda Mandelstam, *Hope Abandoned* (Harmondsworth: Penguin, 1976), p. 211.

Joyce, whose *Ulysses* he claimed not to have read before the early 1930s). Nabokov does not belong in any central way to English literary tradition, in terms of inheritance or legacy; he is best placed, if place him we must, with such rich and strange exceptions as the Sterne of *Tristram Shandy* or the Carroll of the *Alice* tales: company he would have found more to his taste, on the whole, than that of the high priests of Anglo-American modernism.

Western European culture (the last item on his list of self-definitions) offers perhaps a more hospitable pigeonhole for Nabokov's wonderbird. The nineteenth-century novel which he admired as profoundly (albeit differently) as he admired *Anna Karenin* was *Madame Bovary*. Flaubert's painstakingly precise artistry doubtless struck a responsive chord; and, in the character of Emma, imaginative energy compromised by vulgarity of vision and sordid appetite may have found a distant echo in Humbert Humbert of *Lolita*. Yet Flaubert cannot really be cited as a fictional forefather. Kafka, with the implacable surreal logic of his vision, seemed an obvious case of influence to many critics, who cited in particular *Invitation to a Beheading* and *Bend Sinister* (to be considered in the following chapter); there are assuredly affinities between the two writers, and Nabokov acknowledged Kafka's greatness, but insisted that at the time of writing the earlier novel, he had not read Kafka because his German was not good enough, and Kafka had not, at that stage, been translated (a similar situation recurred in the case of the Argentinian writer, Jorge Luis Borges, whose erudition in recondite areas of literature and playful fabulation reminded some of aspects of Nabokov). As twentieth-century history rolled on, and it became unfashionable to admire the Soviet system and fashionable to admire the writers persecuted under communism, Nabokov could have been presented as an *ur*-dissident; but the absence of an obvious political orientation in his work makes this problematic. Amongst those who were identified as 'dissident' in the 1970s, there are some similarities between Nabokov and the Czech exile Milan Kundera, whose novels have a pronounced sexual content and an overt playfulness: but here again the comparison points up as many differences, since Kundera does address himself to politics, and Nabokov is more devoted than Kundera to a vision of the novel as an imperial structure.[7] He shares

7. My comments on Kundera precede the British publication of his latest novel, *Immortality* – his first novel to be written in French, by which Kundera joins Nabokov and Beckett as a writer whose exile is linguistic as well as geographical. Kundera's interest in the concept of 'kitsch' has some points of similarity with Nabokov's interest in *'poshlust'*, which as defined in his study of Gogol is a sort of swooning vulgarity operating as a principle of taste.

with many of these writers, however, a sense of the importance of art's playfulness as identifying an area of the human spirit profoundly hostile to totalitarian systems (and dictators have, of course, acknowledged both importance and hostility in their regular imprisonings of awkward artists).

Asked to describe an Egyptian crocodile to a Roman who had never seen one, Antony jestingly replies that 'It is shap'd, sir, like itself, and it is as broad as it hath breadth; it is just so high as it is, and moves with it own organs' (*Antony and Cleopatra* II vii). A similarly self-referential definition might be best for Nabokov, whom we cannot satisfactorily assign to any easy category; he would, I think, be perfectly happy to be left as a nonpareil or nonesuch, uncageable by any zoo in literary criticism. Yet in the *Wunderkammer* that is literature itself he is effortlessly accommodated, a significant innovator who explored and expanded the possibilities of fiction. There is much to be said in favour of Nabokov's distrust of literary study's conventional classification by movements and schools; his sense of the particular work of the particular writer being the locus for the excitements of reading (and the analytical thrill of re-reading) is a valuable corrective to such abstract overviews. Whatever his intolerance of reputations he considered (with or without justice) to be puffed up, and whatever impatient *hauteur* he could project toward less gifted practitioners, his truer sense was that good writers and good readers alike are freemen of the republic of letters: a more enduring citizenship than any other. Our difficulty with his literary passport is the mark of his 'extraterritorial' status (the term George Steiner applied to him), and however we resolve that difficulty the novels remain. The object of the following chapters is to examine particular novels; my object in this has been to offer the background helpful to a reading of Nabokov's fiction: his childhood and expatriation, his multi-lingualism, his lepidoptery and his chess, his precision and his playfulness. Whilst he might have agreed with George Santayana's declaration that 'the imagination is the true realm of man's infinity', he also knew that 'imagination without knowledge leads no farther than the back yard of primitive art, the child's scrawl on the fence, and the crank's message in the marketplace' (SO p. 32). As he put it in *Speak, Memory*, 'to try to express one's position in regard to the universe embraced by consciousness, is an immemorial urge'; that emphasis on 'consciousness' is significant, for Nabokov's art is more a matter of lucid calculation than of inspirational leaps.

2

The Particular and the General

Invitation to a Beheading and *Bend Sinister* are the two of Nabokov's novels which come closest to exhibiting a 'political' aspect, an authorial comment on modern history: for both unfold in the baleful setting of a totalitarian régime, and each explores the pressures brought to bear by such a system on the would-be dissenter. *Invitation* (as it will hereafter be abbreviated), the eighth of his novels, was composed in Berlin and first published serially in Paris (1935–6) in a magazine devoted to émigré Russian writing. It was issued as a book in 1938, again in Paris; by which time Nabokov and his family were living there, having left Berlin to Hitler. *Invitation* was translated into English from the Russian by Dmitri Nabokov, and published in 1959 (US) and 1960 (UK). *Bend Sinister* was written at the end of World War II in Cambridge, Massachusetts, whither the Nabokovs had transplanted themselves, once more, shortly before the fall of France in 1940. First published in 1947 (US) and 1960 (UK), it was Nabokov's second novel written in English, and his eleventh altogether.

Chased by Bolshevism from his homeland, and then by Nazism from Berlin and Paris, when writing both these novels Nabokov had had the chance to take an uncomfortably close look at the methods of one-party states; and it is natural to imagine that his fiction would reflect what he saw, and what he felt about it. In the foreword to his son's translation of *Invitation*, however, Nabokov cautions 'the good reader' against supposing that the author's attitude toward these political upheavals is an important part of that book. Similarly, four years later in his foreword to the reissue of *Bend Sinister*, whose 'obvious affinities' with the earlier novel is noted, he asserts that 'the influence of my epoch on my present book is as negligible as the influence of my books, or at least of this book [which had sold poorly at first publication] on my epoch'; he is anxious that no one should suppose it to be an example of ' "serious literature" (which is a euphemism for the hollow profundity and the ever-welcome commonplace)'. Neither book inhabits the realm of general ideas, about politics, about society, about anything at all, their author insists; and the energy with which he disclaims such intentions, in these forewords and elsewhere, is matched only by the vigour with which he rebuts the imputation that the writing betrays literary influence – specifically, that of Franz Kafka or George Orwell: 'automatic comparison between *Bend Sinister* and Kafka's creations or Orwell's clichés would go merely to prove that the automaton could not have read either the great German writer or the mediocre English one'. It is of no large significance whether or not Nabokov was influenced by Kafka, since actual plagiarism is not in question. Of greater interest is his objection to the notion that he had been: it was important for Nabokov to be considered an original, and he felt keenly that to perceive one writer in terms of another was to begin a kind of intellectual regression towards the mean, whereby, if you looked loosely and perniciously enough, every writer was pretty much alike in that they all tended to deal with the same subject matter, *e.g.* 'Life', 'Love', 'Death', and 'the Human Condition'. In other words, it was the first step toward considering, not *how* a writer operated in the specifics of his art, but instead the themes and general ideas which could be condensed from it; and for Nabokov, a writer existed only in the particulars of his performance.

Given that in Nabokov's view the genre of 'political novel' hurries its reader toward the literary perdition of social analysis and resonant message, it is unsurprising that he wished to disentangle *Invitation* and *Bend Sinister* from that category. This did not mean, however, that he entirely disavowed a political orientation or political consequence for his work: in a letter he wrote (in Russian) to fellow-novelist Alexander Solzhenitsyn in 1974, congratulating him on his 'passage' from the 'dreadful' Soviet Union of the Brezhnev era, Nabokov mentioned his own novels such as *Invitation* 'and especially *Bend Sinister*, in which, ever since the vile times of Lenin, I have not ceased to mock the philistinism of Sovietized Russia and to thunder against the very kind of vicious cruelty of which you write and of which you will now write freely';[1] (this is an interesting instance in Nabokov's life, of a motif encountered in his art: that of greeting a person who has made the transition between different states of being). Nabokov goes on in

1. *Vladimir Nabokov: Selected Letters 1940–1977*, eds. Dmitri Nabokov and Matthew Bruccolli (London: Weidenfeld and Nicolson, 1990), p. 528. Hereafter cited in the running text thus: (L p. 528).

this letter to affirm that he does not make 'public' political statements; and it is clear that a world of difference exists between these novels and, say, *One Day in the Life of Ivan Denisovich* or *Cancer Ward*. In fact, Nabokov did not greatly care for Solzhenitsyn's fiction, however much he admired that writer's courage. But if he felt that *Bend Sinister* made a protest against Soviet inhumanity, it did not do so by documenting its monstrosities in the way that Solzhenitsyn did, nor by offering an easily-interpretable political allegory like Orwell's; it did so by itself exemplifying the very forms of life and thought such a system cannot accommodate, and typically seeks to eliminate.

Invitation was one of Nabokov's favourites amongst his Russian works. Its setting is an obviously make-believe land, a sort of Ruritania with torture-chambers, in regression from a technologically and morally superior past of which piecemeal evidence remains (an aeroplane or two; glossy magazines). The tale opens in a courtroom, but this owes less to Kafka than to Lewis Carroll; we might think of the trial of the Knave of Hearts in *Alice in Wonderland* (which in 1923 Nabokov had translated into Russian). For *Invitation* generates, not the remorseless nightmare logic of Joseph K's predicament in *The Trial*, so much as the uneasy dreamishness that Alice moves through: with the crucial difference that here, when the cry goes up 'Off with his head!', there is an executioner on hand to do it, and the sense of a life to be forfeited. Yet despite dealing with a death-sentence, the novel has a sustained digressive playfulness in its very texture, a willingness to pull away from conventions of realism and of narrational logic, a continual divergence from straight-forwardness. It does so, not merely as evidence of authorial *jeux d'esprit*, but because this exemplifies the nonconformist modes of consciousness embodied in its central character – which can, if he lets them, save his life. The very first sentence pronounces his death-sentence, and the last scene finds him on the scaffold, but the foreseen plot has an unforeseen unknotting: the deadliest sentence, in Nabokov's fiction, is one whose grammatical and semantic sequence is predictably fulfilled, a kind of cliché; the truly living sentence surprises with an unexpected shift, whether of language, imagery, or meaning.

The plot, such as it is, revolves around Cincinnatus C., condemned to death for a crime which is not identified until Chapter Six, and turns out to be that of 'gnostical turpitude'. This is a crime of being, not of doing: from his childhood on sensitive Cincinnatus had been aware of a property of his selfhood which it was best to conceal; that he was 'impervious to the rays of others, and therefore produced when off his guard a bizarre impression, as of a lone dark obstacle in this world of souls transparent to one another'. Although he has learnt to manage by feigning 'translucence' so as not to alienate his playmates, as he ages he more frequently lets drop his guard – to the perturbation of his fellow citizens, who numerously inform on him:

> Cincinnatus, who seemed pitch-black to them, as though he had been cut out of a cord-size block of night, opaque Cincinnatus would turn this way and that, trying to catch the rays, trying with desperate haste to stand in such a way as to seem translucent. Those around him understood each other at the first word, since they had no words that would end in an unexpected way, perhaps in some archaic letter, an upsilamba, becoming a bird or a catapult with wondrous consequences. (p. 23)

His condition is associated with wondrousness, and with an appetite for the extraordinary and unforeseen (he loves books, and unlike everyone else is fascinated by the 'collection of rare, marvellous objects' in the town museum); he has, in other words, both curiosity and imagination, and therein offers the most terrible affront to the principles of ordinariness governing the see-through lives around him. His predicament resembles that of the protagonist of H. G. Wells's story 'The Country of the Blind', who chances upon a society of the sightless, shortly to find that their proposed remedy for his infringements of their ways of doing things is to put out his eyes.

Cincinnatus C. is not a rebel; he is a gentle individual who has no choice in his nonconformity. His very name (which in Latin corresponds to 'Curly') is exotic, and the repeated 'c' perhaps suggests a man who 'sees' too much (small wonder he is described as 'seasick' on first being incarcerated!). The unintentionality of his transgression is no protection in his dunderheaded town, however, whose hostility to visionary states extends even to the attempt to legislate for behaviour in dreams, as in rule 6 displayed on his cell-wall:

> It is desirable that the inmate should not have at all, or, if he does, should immediately himself suppress nocturnal dreams whose content might be incompatible with the condition and status of prisoner, such as: resplendent landscapes, outings with friends, family dinners, as well as sexual intercourse with persons who in real life and in the waking state would not suffer said individual to come near, which individual will therefore be considered by the law to be guilty of rape. (p. 42)

In this burlesque regulation the issue of transparency, the compulsory absence of any secret selfhood, recurs: even dreams are liable to inspection by the authorities (it is appropriate, here, to recall Nabokov's unrelenting contempt for the interpretation of dreams and similar psychoanalytical intrusions). The refusal by these authorities to admit a distinction between conscious actions and dream-deeds is only to be expected, on the parts of those who turn out to be themselves the very stuff of which bad dreams are made.

'Whoso would be a man must be a nonconformist,' declared Emerson (in his essay 'Self-Reliance'), but in the world of *Invitation* to be so is a capital offence, and such a deviant will not be left to his own devices; as Emerson's disciple Thoreau discovered, 'wherever a man goes, men will pursue him and paw him with their dirty institutions, and, if they can, constrain him to belong to their desperate odd-fellow society'.[2] The hideous insecurity of a desperate society when confronted by dissent is apparent in this novel, for it is not enough that Cincinnatus be simply put to death: before he can be executed, he must be brought to comply with their system – in effect, to consent to the order of reality it represents. If they can constrain him to belong, then they can kill him (and capital punishment is, of course, one of the dirtiest institutions of any society). This is really what the book is 'about': whether they can succeed in abolishing his sense of his own uniqueness, which is a subversion of their corporate and unindividuated modes of being; for in this story, inhumankind cannot bear very much reality.

The malignant irreality of the forces arraigned against this solitary prisoner is stressed throughout; as, for example, the first appearance in his cell of the Prison Director, Rodrig Ivanovich:

> He was dressed as always in a frock coat and held himself exquisitely straight, chest out, one hand in his bosom, the other behind his back. A perfect toupee, black as pitch, and with a waxy parting, smoothly covered his head. His face, selected without love, with its thick sallow cheeks and somewhat obsolete system of wrinkles, was enlivened in a sense by two, and only by two, bulging eyes. Moving his legs evenly in columnar trousers, he strode from the wall to the table, almost to the cot – but, in spite of his majestic solidity, he calmly vanished, dissolving into the air. A minute later, however, the door opened once again, this time with the familiar grating sound, and, dressed as always in a frock coat, his chest out, in came the same person. (p. 14)

The instability of this 'person' is such that his character is interchangeable with that of the jailer, Rodion: one is apt to turn into the other during the course of a paragraph. Cincinnatus is aware of the phantasmal nature of his tormentors, but is unable wholly to resist

---

2. This comes from the chapter entitled 'The Village', in *Walden* (1854).

their power: 'I obey you, spectres, werewolves, parodies. I obey you'. Although their world is of the most half-hearted artificiality, his fear of death attaches him to it, and offers them a means of manipulation; they exploit the weakness of his sensitivity, and with childish cruelty delay or deny his wish for an interview with his (spectacularly faithless) wife, or to know the date of his execution. Every arrangement they make corresponds to a mad, supremely vulgar theatricality, rather than to human need.

This has been so from the beginning where, 'in accordance with the law' the judge pronounced the death sentence by whispering in the condemned man's ear: a singularly loathsome and parodic conjunction of the public and the intimate. The phrase he actually utters – '"with the gracious consent of the audience, you will be made to don the red tophat"' – combining coy locutions with a sense of spectacle, similarly evokes the corruptest kind of amateur dramatics, in which illusion and actuality are yoked by violence together, and real blood mingles with the greasepaint. Nabokov may have had in mind, distantly, those despicable performances mounted in the declining Roman empire, *tableaux vivants* – or, more accurately, *tableaux mourants* – in which death was actually undergone on-stage by condemned criminals, producing for the audience 'that complete pleasure which "art" shot through with "human interest" is said to produce' (*Bend Sinister*, p. 133). He may also have remembered Gogol's friend Zhukovsky, a poet who had evolved a not-much-less disgusting notion that capital punishment would best be undertaken as 'a religious mystery with the hanging performed in a closed church-like place to the elevated sound of hymns, all this invisible to the kneeling crowd, but auditorially very beautiful and solemn and inspiring,' so that 'the enclosure, the curtains, the rich voices of the clergy and the choir (drowning any unseemly sound) would "prevent the condemned man from treating onlookers to a sinful display of swaggering and pluck in the face of death"'.[3] A third sorry item in this list of barbarous theatricals, for Nabokov, could well have been the mock executions to which dissidents were subjected in Tsarist Russia – among them, Dostoevsky.[4]

There are, to retain our theatrical metaphor, three potential scripts for Cincinnatus to follow: one of the authorities' devising; his own wishful projects of escape from the fortress; and the triumphant

---

3. This comes from Nabokov's study *Nikolai Gogol* (New York: New Directions, 1944, 'corrected edition' 1961), pp. 28–9.
4. See Nabokov's essay on Dostoevsky in his *Lectures on Russian Literature* (details in my Bibliography).

one of the author's devising. The first demands that he become as inauthentic as everyone around him. It appears to be the case that, just as the law requires counsels for the defence and prosecution to be uterine brothers, by the same idiotic symmetrification of bad art an executioner (known as a 'fate-mate') is required to establish a bond of brotherhood ('*Bruderschaft*') with his designated victim, more securely attaching him to the things of this world in order more keenly to savour the thrill of severing him from them. To this end the egregious M'sieur Pierre is installed in the next-door cell, posing as a fellow-prisoner but from the beginning (so bad are the authorities at their play-acting) marked out by deferential treatment from the prison functionaries. At every opportunity he imposes himself on Cincinnatus, who is not long deceived by his pretence of camaraderie, and guesses his true identity just before it is revealed. Plump and smelly, M'sieur Pierre lays stress on his own physical prowess, and indeed is rooted in his body – unlike Cincinnatus, who at one stage takes his off entirely, and who remembers walking on air, as a boy. The highest point of M'sieur Pierre's achievement occurs on the eve-of-execution supper to which, 'dressed in identical capes' under which they are 'identically clad', custom dictated headsman and condemned should go. After the meal, during which he had solicitously plied Cincinnatus with unwanted victuals, the two of them observe from a terrace in the Tamara Gardens (of which Cincinnatus had had fond memories), a spectacle breathtakingly jejune: a million or so multi-coloured light-bulbs have been distributed 'in such a way as to embrace the whole nocturnal landscape with a grandiose monogram of 'P' and 'C', which, however, had not quite come off'. This feat of technical wizardry is sustained for a good three minutes and, moved by the lovely sight, M'sieur Pierre crowns his barely-controlled homoerotic frenzy toward Cincinnatus, by pressing his cheek against his victim's.

The second script, that of Cincinnatus's own hopes and fears, resembles the conventional escape narrative in which a prisoner eludes his jailers. What happens here parodies the devices by which this might occur: he tries to gain the confidence of the Director's young daughter Emmie, and she proposes a plan of escape; but what to him is in deadly earnest is to her a game, and when she meets him outside the fortress walls she leads him straight back to his captors. There is, of course, a tunnel; Cincinnatus hears its distant excavation coming nearer, louder, expanding his hopes until his wall crashes in and out pop M'sieur Pierre and Rodrig Ivanovich, dressed up as clowns; like Emmie, treating as material for a practical joke what is for him a matter of life and death. These circular excursions, which lead him from his cell straight back inside again, are a means of demonstrating that his problems cannot be solved on the level he is addressing it. Where is freedom to be found, where could he live? The long looked forward to interview with his wife Marthe (neither the first nor the last semi-human traitress to appear in Nabokov's fiction) is a farce, with every member of the family in attendance, together with their personal effects – and defects, as well. Her unabashed and wholly unrestrained promiscuity underlines the fact that, like all the others, she has no self to betray; she justifies her behaviour by observing to her husband that she is a 'kind creature', as well as that 'it's such a small thing, and it's such a relief to a man'. Small-minded and devoid of compassion, she well reflects a community which – as beheader and beheadee pass excited townsfolk, on their way to Thriller Square with its red-draped scaffold – Nabokov depicts as mounting toward a gratifying collective spasm of murderous bad taste:

> One small house was especially well decked out: its door opened quickly, a youth came out, and his entire family followed to see him off – this day he had reached execution-attending age; mother was smiling through her tears, granny was thrusting a sandwich into his knapsack, kid brother was handing him his staff. (p. 186)

Even while, sick with fear, he rides beside his unctuous 'fate-mate' on this final journey, Cincinnatus is aware that the illusion of the town and its inhabitants actually existing, progressively loses force: trees topple over, the paintedness of the backdrop grows ever more apparent, people are flagrantly two-dimensional. This is the third script, the unconventional narrative that challenges the set text of crime and punishment with its killingly predictable outcome, and overturns the implacable patterning of the death sentence. It also shows him his error in hoping to escape from jail: for in trying to escape, he had consented to the fact of his imprisonment. His problem, all along, is that he has been attempting to live in the wrong world. 'What a man thinks of himself, that it is which determines, or rather indicates, his fate,' declared Thoreau; and once Cincinnatus fully understands himself, the creaking mechanism of terror becomes unsubstantial. The novel ends, therefore, not with him losing his life, but with 'life' draining away from the spectral world around him; at the very last moment, he decides to decline his invitation to be beheaded.

Although the ending has been prepared for by numerous details during the course of narration, it is surprising, because it solves the problem on a plane other than that on which the reader had supposed it to exist. It was not an 'either/or' solution. We might think of Edgar Allan Poe's tale, 'Murders in the Rue Morgue', which looks like an early form of the 'whodunnit', but in fact defeats its genre because the two women in question turn out to have been killed by an orang-utang. This means that there is no 'guilty party', since orang-utangs probably don't feel guilt and certainly aren't legally accountable; and this in turn means that, despite the corpses, no murder has been committed, no trial can ensue, and no punishment can be meted out. To borrow Nabokov's own figure, the tale has taken a 'knight's move' into another fictional dimension; and something similar happens at the end of *Invitation*. The way in which Cincinnatus gains stature as the threatening world about him decays and diminishes is, to be sure, reminiscent of the end of *Alice in Wonderland*, where she grows too large for the courtroom and the playing-cards fly helplessly against her; then she wakes up, and the events of the preceding narrative are relegated to the status of dream. Nabokov's novel ends with Cincinnatus moving 'in that direction where, to judge by the voices, stood beings akin to him'; where can this be – *our* world, 'Heaven'? What does seem likely, is that the story has not been a third-person authorial narration on Nabokov's part, but is supposed to have been recounted by Cincinnatus himself, looking back from his new vantage-point (this seems to be indicated by the unexpected vocative use of 'you' on p. 116, 1. 10, and p. 146, 1. 19); it is, in other words, the story of a man who has come through, and who can look back (as perhaps Nabokov, in 1974, supposed Solzhenitsyn would) on a nightmarish previous existence.

Nabokov's book on Gogol (1944) is notable, among other things, for starting with that writer's death and ending with the date of his birth. We have seen that *Invitation* begins with the death sentence, and ends with Cincinnatus's rebirth as his true self. A not dissimilar pattern is discernible in *Bend Sinister*, which commences with the death of Krug's wife, Olga, and closes with his annihilating intimation of a higher plane of being than his own. Its concerns have some similarities with those of the earlier novel: there, the subject was 'the precious quality of Cincinnatus', his human spiritual quiddity; here, the author's foreword identifies the book's 'main theme' as 'the beating of Krug's loving heart, the torture an intense tenderness is subjected to'. Although, as has already been noted, Nabokov disclaimed for it the status of 'serious literature', it is a more serious book than *Invitation*; both concern the predicament of an exceptional individual in hostile circumstances, but in *Bend Sinister* the forces ranged against Krug are not so easily dismissed as, finally, those ranged against Cincinnatus are. Although in each novel we observe the conflict between what is unique and unaffiliated, and what is collective and incorporated, in

Bend Sinister the small-town stupidity and fear of the exceptional, seen in the earlier novel, have spawned a dictatorship whose functionaries are less easy to laugh away than M'sieur Pierre and his minions. A state of mind has here established itself as a police state, and it is for Krug more than a matter of waking up.

In his foreword Nabokov looks back on his life during the novel's composition (1945–6) as having been 'particularly cloudless and vigorous' – although in fact he nearly had a nervous breakdown in the summer of 1946 (see NWL p. 170). Notwithstanding the happinesses the foreword enumerates, it may be that the darker tones of *Bend Sinister* owe something to its author's augmented sense of the hideousness of recent history. In a letter to his sister Elena of 26 November 1945, he likened the bureaucratic obstacles that had hindered his family's escape from France in 1940 to something that might have occurred in *Invitation*; but in the same letter he alluded to their brother Sergey, who had not fled and – Nabokov had recently learnt – had died in a Nazi concentration camp (Neuengamme), where he had been interned because of his homosexuality. In this the hideousness of history became particular, not general; and happy as he was in his adopted country, Nabokov must have reflected on how close he and his family also came to disaster: his wife, Véra, was Jewish, and she and their son Dmitri would have been at enormous risk under the Nazis, only one step behind them when they sailed away. *Invitation* and *Bend Sinister* have in common that they depict the persecution of a superior being by a society of dolts; but in the later novel a more candid assessment of the danger these dolts pose is made, and there is a keener awareness of the vulnerability even of the being who realises his own superiority. Unlike Cincinnatus, who can at last shrug aside his ghastly 'family', Krug has a beloved child; questions of loyalty and betrayal are more sharply posed; but at the same time, the novel's verbal inventiveness and technical extravagances outstrip those of *Invitation*.

The story concerns Adam Krug, a philosopher of international reputation in a small unidentified country with a middle-European flavour. It opens with his dissociated but specific observations of things noticed (puddle, trees, houses) beyond the window of the hospital room in which he sits, attempting to come to terms with the unexpected news that his wife will die, after an unsuccessful kidney operation. The morose, insulated quality of his mental interior is due in part to his grief, in part to the pint of brandy which we learn he has consumed to cope with it. He and Olga had been very much in love, and in addition to his own numbed shock Krug has to face the problem of how he can explain her lengthening unavailability to their unsuspecting son, David. There is on top of this an intellectual problem in the event, since it appears that Krug cannot be reconciled to the fact of death, and refuses to acknowledge it: he will not speak of his dead wife to the friends whom, in his defiant absence, he leaves to organise the cremation of the unspeakable and in his eyes irrelevant evidence that death exists.

Olga Krug's decline has coincided with a fateful sickness in the body politic: whose symptoms are the surly and officious soldiers imposing a curfew, and whose source is the newly-installed dictatorship of Paduk. The inner turmoil of the widowed husband, his strong mind virtually disabled by grief, thus takes place against a backdrop of political upheaval; and the novel's 'plot' concerns the interactions and collisions between Krug's private, intensely personal preoccupations, and the public pressures which his very eminence creates for him. Both aspects distinguish this book from *Invitation*: for whereas Cincinnatus, in order to resolve his problem, had simply to understand that he was realer than the phantoms which afflicted him, however much Adam Krug may despise politics and its practitioners, he lives in the fallen world (suggested by his name, as well as several references to apples), in a political ordering which is as lethal as it is laughable. Although for much of the book he supposes he may decline its invitation, he finally discovers what its powers amount to. As will be seen, in its writing *Bend Sinister* is in some respects more playful than *Invitation*; but also, semi-paradoxically, it is to a far greater extent involved in questions of 'human interest'. Cincinnatus C. is something of a cipher, whose very name resonates with its own unlikelihood; but the portrayal of Adam Krug is of a much denser human being: to be effective, the novel must make the reader participate in his past love for Olga and his present love for David.

In both his powerful love and his capacity for original thought, Krug represents qualities inimical to the police state in which he finds himself; whose leader has founded the 'Party of the Average Man', with all that implies about its attitude toward an exceptional man such as Krug. Insofar as it has a political philosophy, this derives from the theories of one Fradrik Skotoma ('scotoma' is a blind spot in the eye) concerning the achievement of human happiness by the equal redistribution, not of wealth, but of 'human consciousness', in whose uneven apportionment amongst mankind 'lay the root of all our woes' (a cliché appropriate to this drab thinker's style). In Skotoma's vision (or blindness), 'human beings . . . were so many vessels containing unequal portions of this essentially uniform consciousness,' and to counteract this inequality he proposed 'the idea of balance as a basis for universal bliss and called his theory "Ekwilism"'; ('Ekwilism' is presumably a distortion of 'equalism').

Skotoma was innocent of envisaging the means by which this might be achieved, and merely asserted 'that the difference between the proudest intellect and the humblest stupidity depended entirely upon the degree of "world consciousness" condensed in this or that individual', relying on the evident truth of that proposition to effect the desired change; however addle-pated, his book had been intended to produce human happiness. Dead by the time Paduk took it as the basic text for his Party, Skotoma had been 'spared the discomfort of seeing his vague and benevolent Ekwilism transformed (while retaining its name) into a violent and virulent political doctrine', whose end was 'to enforce spiritual uniformity upon his native land through the medium of the most standardized section of the inhabitants, namely the Army' (p. 71).

The figure of Skotoma may owe something to Nikolay Chernyshevsky (1828–89), exiled by the Tsar but admired by Lenin; his practice and theories laid the foundations for the doctrine of 'socialist realism', so balefully applied in the Soviet Union (his biography as written by the hero, Fyodor, forms a substantial part of *The Gift*). In any case, in *Bend Sinister* Nabokov offers rather a broad satire of communism and its discontents; although not so broad as to be unrelated to the historical actuality whereby, in Soviet Russia and elsewhere, a political structure based on theories of fairness and freedom ended by systematically disenfranchising and terrorising its constituents, exterminating millions but still finding itself unable adequately to feed or house those left alive. Facts of that magnitude can hardly be well addressed in fiction, although some have tried; Nabokov's method in *Bend Sinister* is to focus on the consequences for the individual, Krug, of Paduk's version of Ekwilism with its sinister tendencies toward the general good; and if there is a palpable design on the author's part, it is – as he implies in the foreword – that the reader should respond to Krug's capacity for humanity, rather than to Paduk's capacity for inhumanity.

In this, Nabokov sets himself a somewhat harder task than he had in the earlier novel. Cincinnatus was a gentle – if also somewhat notional – soul, beguilingly endowed with attributes of loyalty, fear, trustingness, and sensitivity; Professor Krug, on the other hand, is rebarbative and arrogant, although to some degree these manifestations are caused by the fact that he is a wounded animal: 'a great big sad hog of a man', as he describes himself. Unlike the air-walking Cincinnatus, Krug's physicality and even clumsiness is emphasised: he will not find it easy to shake off his body, and in this book it is the dead Olga who, in an image reminiscent of one in *Invitation*, is imagined seated at her dressing-table, removing her jewellery, and along with it the components of her body bit by bit, in her husband's queasily erotic dream. His brusqueness with his servant Claudina, his boorishness toward the frightened old refugee who cowers in the not-working lift of Krug's apartment block, are unendearing features in the portrait; and even more so is the unworldly stupidity (perhaps another indicator of his brilliant intellect) with which he refuses to confront the perils of his situation.

People have uses for Krug: his University would like him to intercede on its behalf with the new Dictator, presuming that their schooldays together will have formed an exploitable bond (in fact, Krug bullied Paduk); the régime itself, anxious for international respectability, would like to instal him at the head of the University, to proclaim that all was well. Agents are set to work to probe for his 'handle'. For his part Krug persists with his aloofness: 'I am not interested in politics' he declares, when warned about checkpoints in the city, but nevertheless the goonish soldiers send him back and forth across the bridge they guard; when called to an interview with Paduk, Krug informs him that 'I am not in the least interested in your government. What I resent is your attempt to make me interested in it. Leave me alone'. This echoes his earlier protestation to the prosaic but fiercely well-meaning Maximov, that he wanted 'to be left alone'; to which Maximov's retort was 'You are not alone! You have a child'. This argument with Maximov in Chapter 6 is one aspect of the greater depth of *Bend Sinister*, for it carries its own critique of Krug, as well as a foretaste of his disaster. Walking back to the cottage with his son later that day, Krug ungenerously reflects on Maximov, who had implored him to flee the country, as a 'fool', and assures himself, 'excuse me, I am invulnerable'. But he returns to find the Maximovs have been spirited away by the secret police; and he holds his vulnerability by its childish hand.

One of Krug's professorial colleagues observes to another that 'curiosity . . . is insubordination in its purest form'. It is this quality, the free play of the questing mind, which Nabokov associates with childhood at its best: he remembers it from his own experience, he observes it in his son Dmitri (in *Speak Memory*), and he endows Krug's son with it. Such curiosity is a receptive interest in the surrounding world, a willingness to be intrigued that is accompanied by a disgust for things which do not adequately reward that interest – such as the toy house in the village shop which momentarily engrosses David, but turns out to have painted, not transpicuous, windows. Opposed to this is the toy car he has just been bought, which pleases the child's eye for detail by having realistic rubber tyres (this is a memory of Dmitri); only stupid adults suppose that children are easily fooled, and the ability to sympathise with a child's universe is associated by Nabokov with the capacity to respond to detail: the novel scathingly dismisses the sort of person who cannot notice 'the wallpaper in a chance room', or is incapable of 'talking intelligently to a child'. Childhood itself, with its undisciplined energies, is gross insubordination to a conformist state; and the immoderation of a father's love for this particular soul refutes the 'average' response:

> And what agony, thought Krug the thinker, to love so madly a little creature, formed in some mysterious fashion . . . by the fusion of two mysteries or rather two sets of a trillion of mysteries each; formed by a fusion which is, at the same time, a matter of choice and a matter of chance and a matter of pure enchantment; thus formed and then permitted to accumulate trillions of its own mysteries; the whole suffused with consciousness, which is the only real thing in the world and the greatest mystery of all. (p. 156)

David represents the vibrating individuality of consciousness, whereas the Ekwilist regards 'consciousness' as a generalized phenomenon. When in 1944 Nabokov started work on the novel, its provisional title was 'The Person from Porlock' – that classic interrupter of a flight of fancy. His ghost lingers in the first and last letters of Paduk's name, who is a more determined enemy of the imagination. For example, he brings to school a machine of his father's invention, the 'padograph', which is a sort of typewriter of handwriting, custom-made to reproduce mechanically the irregularities of a person's script. The delight of this, from an Ekwilist's point of view, was that the padograph was 'proof of the fact that a mechanical device can reproduce personality, and that Quality is merely the distribution aspect of Quantity'. Paduk's public image, as he rises like Hitler from fatuous obscurity to total power, is modelled on a popular cartoon character, Mr Etermon ('Everyman'), a two-dimensional figment remarkable only for being utterly predictable. Nabokov spends some time amusing himself with a mocking description of Etermon's 'life' as depicted in the cartoons, before denouncing the generalizing habits of perception that underlie such caricature, and paying tribute to the infinite mysteriousness of individuals:

> Actually, with a little perspicacity, one might learn many curious things about Etermons, things that made them so different from one another that Etermon, except as a cartoonist's transient character, could not be said to exist. All of a sudden transfigured, his eyes narrowly glowing, Mr Etermon (whom we have just seen mildly pottering about the house) locks himself up in the bathroom with

his prize – a prize we prefer not to name; another Etermon, straight from his shabby office, slips into the silence of a great library to gloat over certain old maps of which he will not speak at home; a third Etermon with a fourth Etermon's wife anxiously discusses the future of a child she has managed to bear him in secret during the time her husband (now back in his armchair at home) was fighting in a remote jungle where, in his turn, he has seen moths the size of a spread fan, and trees at night pulsating rhythmically with countless fireflies. No, the average vessels are not so simple as they appear: it is a conjuror's set and nobody, not even the enchanter himself, really knows what and how much they hold. (pp. 73–4)

A regimented society which perceives its members *en masse* could only be threatened by conceiving them, not as clones, but as fantastically different from each other; and it follows that childhood, as the germinal state of such anarchic individuality, would be repressed by such a system. If David, full of trustful wonder, represents an ideal of childhood (albeit that he seems a couple of years younger than his age of eight), it should be remembered that the obverse of this is the casual sadism and wilful refusal to feel – except through experiment on others – that are also, frighteningly, childish. The people in positions of authority, both in *Invitation* and *Bend Sinister*, have the attributes of brutal children, frightened of the burden that adult individuation would impose on them, and retreating from it to the conformist security of gangs. These regressive types, when in power, naturally have a world-view in which 'a person who has never belonged to a Masonic Lodge or to a fraternity, club, union, or the like, is an abnormal and dangerous person' (this is almost the exact reverse of Nabokov's own view of the matter). These novels hint at what it might be like to endure the capricious, irresponsible and above all envious tyranny of underdeveloped children; a nightmare with which, alas, 'reality' makes us familiar: in the régime of the Khmer Rouge in Cambodia for example (see David Puttnam's film *The Killing Fields*), or that of the Ceausescus in Roumania.

One by one Krug's friends are whisked away, and he begins to think better of his intention of staying put; but it is when, too late, he attempts to organise an escape for himself and David through a stool-pigeon, that his enemies understand that David is the 'handle'. Nabokov admits in his foreword that a 'real-life' dictatorship would hardly have taken so long to realise this; but the point may be that those who themselves have had no childhood, and either have no children (Paduk is homosexual) or do not value those they have, are not well placed to imagine a father's love. Having at last grasped this fact, however, the authorities arrest Krug and remove his son, intending to blackmail him into co-operation; but due to the inefficiency which is their trademark they mistakenly despatch David to an experimental 'institute', in which unwanted or otherwise inconvenient 'orphans' are made available to adolescent psychopaths for therapeutic dismemberment – the whole process being recorded on film in the interests of 'science'. The kind of film produced is intimated by Nabokov with the blackest of humour; and were it not for the fact that footage actually exists of an obedient Nazi soldier clubbing with his rifle-butt the heads of tottering children (in an 'experiment' to assess the effect of brain-damage on physical co-ordination), we would be inclined to dismiss it as the wildest fictional invention.

The effect of his son's murder on Krug, and the way the novel ends, may be left aside for the moment. For to offer this summary of events substantially misrepresents *Bend Sinister* as a more straightforward act of narration than it is: as Nabokov counsels in his foreword, the story 'is not really about life and death in a grotesque police state'. The reader who is familiar with it will already have noticed how much my account, so far, has left aside: the divagation into Shakespeare and *Hamlet*, for example, introduced through Krug's friend Ember in Chapter Seven; the reminiscence of Olga with a hawk-moth in Chapter Nine; the glimpses we have in other chapters of the kind of book about time and consciousness which Krug wishes to write, and on which he embarks the night he and David (with whom he has quarrelled at bedtime) are separated for ever. *Bend Sinister* is essentially itself, in its deviations from a straightforward 'human interest' story about Krug and his son; its subject-matter has to do with conformity and nonconformity, and therefore it is desirable to register those ways in which the novel consistently refuses to conform to the narrative expectations we bring to it. *Laughter in the Dark* (written in Russian, 1932) gives a summary of its own plot in the short first paragraph, and then notes that 'although there is plenty of space on a grave-stone to contain, bound in moss, the abridged version of a man's life, detail is always welcome'; it is in the detail, the specifics of a text (as of a life), that art operates and interest lies.

We all are born and we all die; but it is in those aspects of our lives which elude such crashing generalizations that our individuality declares itself, as 'the accent of deviation in the living thing / That is its life preserved' (in the marvellous words of Wallace Stevens's poem 'A Discovery of Thought'). Nabokov has asserted that 'one of the functions of all my novels is to prove that the novel in general does not exist' (SO p. 115); and one way of appreciating a work like *Bend Sinister* is to think of it as constantly deviating from the pattern of the more conventional novel it might have been: for it is along the detours from functionality's straight road that the best views are found. In the lectures on fiction Nabokov gave at Wellesley and Cornell, he consistently drew his students' attention to how the works under discussion solved (or failed to solve) the problems of narration in original or unexpected ways; we should be aware of similar elements in his own fiction: for he was a writer who refused to believe that prose should be prosaic. It was quite possibly his irritation that so little of his artistry had been observed by readers, that led him in the foreword to *Bend Sinister* to list some of the devices he had planted in the text.

Without repeating his list, there is plenty to observe. In his comments on other writers Nabokov laid some emphasis on their skill – or otherwise – in managing the transition from one character or scene to another: is it an awkward, abrupt change of gear, or is it sinuously accomplished? *Bend Sinister* furnishes some useful examples of the job well done. In Chapter Three, Krug 'phones their friend Ember to inform him of Olga's death, and the locus of narration moves down the wire and into the letter which Ember had been writing when the call came, into Ember's consciousness for a couple of pages, then back to Krug as he puts down the receiver. More elaborate still is the sequence of Chapters Four, Five, and Six. Chapter Four has mostly been concerned with the sinister meeting at the University, attempting to persuade Krug to bear a message of co-operation to Paduk (nicknamed 'the Toad', from 'paddock'); it ends with an extraordinary paragraph, in which Krug's aching sense of his wife's absence combines with the sequins glimpsed on the cloak of a girl necking with her boyfriend on his doorstep, which sequins suggest (a) the mythologized stars and (via Pascal) the vast emptiness of Olga-less space, and (b) the costume of a trapeze artist (thought), defying gravity as with his mate he performs aerial arabesques whose beauty conceals their effort, before dropping into the circus-net mythology provides for him, to acknowledge the audience's applause and to wipe away his sweat at death defied once more. This highly elaborated closing image prepares the scene and tone of the next chapter, which is a dream Krug has, consisting of memories of schooldays with Paduk and others, symphonically intertwined with a recurring image of a woman, Olga, in cherry-red velvet before her dressing-table. Like many dreams, this has various levels, some more lucid and explicable than others, and it gives Nabokov the opportunity of providing the reader with necessary facts relating to the shared schooldays of Krug and Paduk, together with the origins of Ekwilism, without resorting to bread-and-butter past-tense narration.

Chapter Five ends with the image of Olga's self-discorporation, when 'with a horrible qualm Krug awoke'; the next chapter starts:

' "We met yesterday,", said the room. "I am the spare bedroom in the Maximovs' *dacha* [country house, cottage]. These are windmills on the wallpaper." "That's right," replied Krug' (Nabokov's parenthesis). This unusual dialogue between an occupant and his bedroom is a way of avoiding a more predictable explanation ('Krug had taken David to the Maximov's place in the country, etc.'); it also enacts something of the strangeness we can feel when returning from an intense dream to our waking environment: Krug's initial disorientation is implied, not stated. There are many noteworthy instances where the language mimics, rather than describes, what the reader needs to understand: in the discussion with Maximov that follows, Krug's reaction to the firm (and sound) advice that he should immediately leave the country is implied in this atmospheric paragraph, which interrupts their verbal interchanges:

> The stove crackled gently, and a square clock with two corn-flowers painted on its white wooden face and no glass rapped out the seconds in pica type. The window attempted a smile. A faint infusion of sunshine spread over the distant hill and brought out with a kind of pointless distinction the little farm and its three pine trees on the opposite slope which seemed to move forward and then to retreat again as the wan sun swooned. (p. 83)

This is a more interesting and more effective way of rendering Krug's heart-sickness at the thought of expatriation than direct description would have been. Indeed, the novel on occasions parodies the sort of set-piece description that might be considered 'realistic'; as when in Chapter Four Adam Krug is physically set before us: his size, hair, face, are presented for inspection, but the portrait is interrupted by the following: 'The brain consisted of water, various chemical compounds and a group of highly specialized fats'. The description then reverts to visible details, the whole being capped by this final sentence, a paragraph in itself: 'Under this lay a dead wife and a sleeping child'. In such ways, Nabokov indicates the different levels of 'reality'.

When Krug picks David up from the kindergarten, David casually lets drop that the truth from which his father has attempted to shield him has been revealed; and the stuck-record prose mimics Krug's panic as he plays for time: "[Billy] said my mother was dead. Look, look, a woman chimney sweep." (These had recently appeared owing to some obscure shift or rift or drift in the economics of the State – and much to the delight of the children.) Krug was silent.' (pp. 137–8). And while this happens, we infer, he has inadvertently walked his son into one of the puddles he had instructed him to avoid. In addition to such linguistic enactments of the sense, there are occasions when the language takes on a sort of life of its own, of no discernible relevance to the matter in hand: one paragraph ("The night had been stormy . . .' p. 84) commences with a little storm of sibilants; another (' "Iron and ice," . . .' p. 97) seems to celebrate the various sounds of 'i'; and most spectacular of all is the small explosion of 'v's in the paragraph that straddles pp. 77–8, although there is still room in the background for a little dance of 'b's, 'm's, and 's's. I am unable to ascribe any 'point' to such displays, other than that they celebrate the potential extravagance of language, straining against the teleology of making sense as the free-thinker rebels against the compulsions of orthodoxy. A similar non-utilitarian excess lies in the various languages which echo in this novel (and more so in some later fictions), expressing (I suppose) the potential multiplicity of ways of putting it. As Thoreau declared, 'I yearn some where to speak without bounds, to make an extra-vagant statement'.

At almost every point *Bend Sinister* looks for the unexpected effect – as, for example, in Krug's interview with Paduk (which its creator singled out for mention in letter of 1944), where 'what happened' is subject to continual restaging (e.g. pp. 123–4). Another aspect of Nabokov's practice is the unforeseen co-ordination of small details; the reader will already have noticed that the comment about wallpaper on p. 73 is picked up on p. 76, and in addition to those Nabokov lists in his foreword, there are the three mentions of the student Phokus (pp. 86, 148, 181), two allusions to Pascal (pp. 59 and 118), and the identity of the stammering schoolfellow (p. 69) revealed on p. 199. There are certainly others, but the best, and savagest, of these surprise co-ordinations occurs near the end of the book. From early on we have been made aware that there is another Professor Krug at the university, and unobtrusive confusions or associations of the two have been a constant feature; the series reaches its grimly comic climax when the authorities, keen to reunite the about-to-be-compliant Adam Krug with David, usher in to him Professor Martin Krug's son, Arvid. The hapless functionaries can hardly bring themselves to believe that the mistake is theirs, not Krug's (and the authorities continue to confuse the two Professors). Then, after a long drive, comes the reunion of Adam with his son: 'The murdered child had a crimson and gold turban around its head; its face was skilfully painted and powdered: a mauve blanket, exquisitely smooth, came up to its chin. What looked like a fluffy piebald toy dog was prettily placed at the foot of the bed' (p. 187). Even here, there is a surprise: the dog turns out to be real, and snaps at Krug when he knocks it aside.

Immediately following this, Krug's frenzy is represented to us in a fractured prose consisting of Russian phrases with parenthetical translation and comment: another of the myriad examples of the text's drawing attention to itself as performance and as mediated utterance, which culminates in the extraordinary ending, in which Krug, awaking from a dream in his prison-cell, is visited by the apparition of his creator, Nabokov (who had previously hinted at his presence behind events); a supernatural soliciting which drives him mad, and insulates him from the final farcical impositions of Paduk's operatives. He is apparently shot as he leaps toward the cringing dictator – in a replay of their schooldays – but the locus of narration abruptly shifts to the room in which Nabokov finishes the novel, where a hawk-moth has just twanged against his window's insect-mesh. His character Krug died happy, he concludes, because at the last he discovered that 'death was but a question of style'.

I doubt that many readers escape a twinge of dissatisfaction with this closure. At the worst, it is rather like coming to the end of what seemed like a good meal (marred slightly by the host's reiterated declarations that he had done all the very complex cooking himself), to be told that in spite of the fact that it all looked like food, there has been no nutritional value whatever in the experience. It is as if Nabokov wishes there to be no act of transference between his fiction and the reader's world; and this is doubly vexatious since, if the reader has done as instructed and attended to the beating of Krug's loving heart, he is likely to feel somewhat cheated at the author's insistence that there was no heart to listen to, all along – or that it was as fake as the amplified heartbeat of Paduk, listened to by pantomime medics.

We may dislike being told that Krug died happily ever after; but Nabokov thought highly of his resolution, regarding it as a significant addition to the practice of fiction. In a letter to a prospective publisher of 22 March 1944, he had this to say of Krug's end: 'to put it bluntly, he realizes suddenly the presence of the Author of things, the Author of him and of his life and of all the lives round him, – the Author is *myself*, the man who writes the book of his life. This singular apotheosis (a device never yet attempted in literature) is, if you like, a kind of symbol of the Divine power' (L pp. 49–50). Considering these comments, it must be remembered that the novel Nabokov described in 1944 differed in some respects – not just in title – from the one he published three years later; we should also bear in mind an author's enthusiasm for a work in progress, as well as his need to 'push' it in such a letter. Nevertheless, the ending he describes resembles the ending as it is, with no obvious distortion; and if it *is* intended to symbolize the power of God, then we confront the possibility of a 'religious' aspect to the novel. I do not find that this helps in understanding why *Bend Sinister* finishes as it does: to affirm the existence of God should presumably console and reassure about the nature of the universe; but the reproach which Krug might level at his creator Nabokov parallels the problem that confronts one asked to believe in the existence of a divine Creator: why, given your total power, have you created evil? Moreover, for most of us the revelation that everything we are has been the consequence of an absolute will external to our own would be horrific, and would deprive our lives of any significant moral dimension by rendering our choices illusory.

It does not seem appropriate to consider this novel in the light of such fundamental religious mysteries, to which it is inevitably inadequate. In the same letter, Nabokov wrote that, despite his distaste for 'message of hope books', he felt that 'a certain very special quality of this book is in itself a kind of justification and redemption, at least in the case of my likes' (L p. 49). These words, too, have a potentially Christian resonance; yet Christianity does not tell us that death is merely 'a question of style'. Our difficulty with the end of *Bend Sinister* is that it seems like an evasion, on more than one level: Krug is absolved from the responsibility of deciding whether or not to sacrifice his imprisoned friends; the author is absolved from taking the evil he has imagined seriously; the pattern is completed, but at some cost. As the foreword warns, the book does not aspire to be 'serious literature': it has throughout exhibited a self-delight in its own powers of artifice and fabrication that is quite independent of an overt moral purpose, and has as constantly stressed the theatricality of its own performance. Nevertheless, however alert we may have been to the hints of authorial stage-management, his sudden intrusion on Krug's end has an apparent flippancy about it. Pascal is evoked in *Bend Sinister,* and we might well recall another of his *Pensées: 'Le dernier acte est sanglant, quelque belle que soit la comédie en tout le reste';*[5] for even in the theatre, the words of Mercury can overpower the songs of Apollo.

Or as Sir Walter Raleigh put it, similarly working a theatrical metaphor to an untheatrical conclusion, 'we dye in earnest, that's no Jest'. This seems diametrically opposed to a view of death as 'a question of style'; but to appreciate possible interconnections between death and style, we need to understand the particular orientations of both *Invitation* and *Bend Sinister* toward matters of morality and matters of art. While he was engaged on the later novel, Nabokov had occasion to set forth his views in a letter (24 October 1945) to a Professor of Russian who had queried his recently-published book on

5. 'The last act is bloody, however fine the rest of the comedy may be'.

Gogol. 'I never meant to deny the moral import of art,' he explained, 'which is certainly inherent in every genuine work of art. What I do deny and am prepared to fight to the last drop of my ink is the deliberate moralizing which to me kills every vestige of art in a work however skilfully written' (L p. 56). The 'justification and redemption' Nabokov discerned in *Bend Sinister* (for him and his kind) is therefore less likely to derive from any quasi-theological perception, than from the operations of what, in his letter to the Professor, he referred to as 'the inherent morality of uninhibited art' (L p. 57).

Seen from this angle, the ways in which his books refuse to conform to type or expectation, the uninhibitedness with which they refute notions of 'the novel in general', are their most moral element, in which the ethics of particularity, of uniqueness, are asserted. 'A foolish consistency,' wrote Emerson, 'is the hobgoblin of little minds'; the larger minds envisaged by Nabokov as his ideal readership thrive on the fictional surprises that he springs, and understand that the creative inconsistencies of his art are of its essence, bringing to birth an imp of the perverse that is in every respect the opposite of Emerson's hobgoblin:

> Deceit, to the point of diabolism, and originality, verging upon the grotesque, were my notions of strategy; and, although in matters of construction I tried to conform, wherever possible, to classical rules, such as economy of force, unity, weeding out of loose ends, I was always ready to sacrifice purity of form to the exigencies of fantastic content, causing form to bulge and burst like a sponge-bag containing a small furious devil. (SM p. 222)

Nabokov here speaks of his composition of chess-problems, but it is clear that these remarks could equally be applied to his composition of novels; and the very way in which this sentence moves from a discussion of technique to the startling final image (who could have foreseen that sponge-bag and its little devil, even with the pointer of 'diabolism'?), exemplifies the liberating irresponsibility of his writing.

Such a sentence offers a deliberate affront to inflexible ideas of order, just as the endings of both *Invitation* and *Bend Sinister* outrage conventional expectations; in doing so they illustrate how art can be free where life is pre-determined. Also, more pragmatically, they illustrate the mental attitude that offers best defence against the totalitarian conditions they portray: a jesting response to tyrannical earnestness. In Nabokov's 1938 short story 'Tyrants Destroyed', the narrator discovers (so it seems) that the most effective way of dealing with the dictator he despises is to laugh at him rather than to assassinate him. Neither Cincinnatus nor Krug accord their persecutors the tribute of seriousness so desperately sought, and thus prevent their power from being absolute. These two are figures of capable imagination (the phrase is Wallace Stevens's) in a world of stunted make-believe, individuals in a generalized society, to which one child is pretty much the same as another, Arvid and David Krug interchangeable quantities. In this view tyrants, addicted as they are to the bad art that bodies forth their grandiose obsessions, are vulnerable to the good art whose precision mocks their vapidity, and whose triumphant nonconformities mock their pathological desire for domination.

It is in this context that death and style can be connected. In his letter to Solzhenitsyn Nabokov defined the major attributes of the Soviet system as 'philistinism' and 'cruelty'; there is little doubt that, although he saw the moral difference between the two conditions, for him one led to the other, and an insensitivity to the mysteries of art could imply an insensitivity to the mysteries of human beings; and an inability to respond sympathetically to another human is a precondition of being able to torture him.[6] The major effect, in post-revolutionary Russia, of Stalin's imposition of the doctrine of 'Socialist Realism' on the arts, was an extreme regressive conservatism in technique operating on a narrowly predictable range of subjects, a 'life-likeness' that in fact was deadliness; both hilarious (to the outsider) and horrifying, in the compulsorily narrowed mental horizons implied.[7] The totalitarian eye-view is always quantitative, statistical rather than detailed ('oh, lots of things!'); its view of human reality is well represented by devices such as the padograph or, in *Invitation*, the photohoroscope, which projects an individual's future on the basis of suitably-adjusted photographs. Its most appreciated art is a matter of simulation, whose culmination is the intermingling of the real with the represented, as in the dog that looks like a toy or, much lower on the scale, the Roman spectacles in which actors really lost their lives.

6. There is a counter-argument to this: that some of the operators of Nazi death-camps were cultured men who appreciated Beethoven. I have not, however, seen convincing evidence of the aesthetic sensitivity of torturers.

7. Igor Golomstock's book *Totalitarian Art* (London: Collins Harvill, 1990), contains some specimens; in particular, a painting called 'The New Flat' by Aleksandr Kaktenov: in which a nuclear family of proletarian party members joyously contemplates (son clutching a portrait of father Stalin to adorn the walls) their new abode with its roominess, books, pot-plant, balalaika and radio. Its 'realism' is historically specious, alas; very few of 'the people' lived to such a standard in Stalin's Russia, and quite a few nuclear families would have had to share those parquet floors. The painting is a good example of what Nabokov called *poshlust* operating in art; this useful term is defined in his study of Gogol.

Nabokov's is an art in which, as we see, people specifically do *not* lose their lives: the executions are false, and Krug's 'death', like Cincinnatus's, is a mirage. It is an art of dissimulation, of successively-revealed layers, having nothing to do with 'simplicity' and 'sincerity', and everything to do with complexity and deception. Its morality is the morality of specific perception (the lepidopterist's eye-view) and untrammelled imagination, celebrating two great human attributes – the ability to look curiously and the ability to dream – in the medium of a great human resource, language, operating in the marvellous, sophisticated human invention of the novel. Nabokov is opposed to didacticism in literature because he saw this as contradicting the profoundest spirit animating it, which is a playfulness that confers temporary liberation from the rules of time and space governing our ordinary living. Art arises out of life, but is an act of insurrection against life's practical imperatives, because of its own sublime uselessness; it is in this aspect that, by its nature, it opposes regulation, and has always been the object of suspicion on the part of tyrants whom, by its nature, it affronts. But it cannot, in Nabokov's view, afford to set itself to direct debate with these 'tigroid monsters, half-witted torturers of man', because to do so would be to destroy its most powerful weapon against them: its disinterestedness.

To engage directly with 'real-life issues' would be like Cincinnatus consenting to the imprisonment from which he then tries to escape. This is why the two novels under discussion sedulously refuse to reach conclusions or offer solutions with regard to the political situations they evoke: it would be self-betrayal. *Invitation to a Beheading* and *Bend Sinister* are not, importantly, 'political novels'; but nor should they, in the light of the foregoing discussion, be seen as somehow aesthetically self-enclosed: remote, irrelevant, smugly aloof. In them and in his other books, Nabokov remains true to an ideal which is firmly based on a sense of the human importance of the values of story-telling; an importance which is given notable expression in some words of Wallace Stevens:

> The final belief is to believe in a fiction, which you know to be a fiction, there being nothing else. The exquisite truth is to know that it is a fiction and that you believe in it willingly.[8]

8. This is one of the sayings collected under the title 'Adagia' in Stevens's *Opus Posthumous*, ed. Samuel French Morse, (London: Faber and Faber, 1959), p. 163.

3

Enchantment and Disenchantment

In one of his letters to Edmund Wilson, Nabokov declared that 'the longer I live the more I become convinced that the only thing that matters in literature, is the (more or less irrational) *shamanstvo* of a book, i.e., that the good writer is first of all an enchanter' (November 1946, NWL p. 177). In the introductory lecture to his course on European fiction at Cornell, Professor Nabokov assured his students year by year that, of the three elements discernible in any major writer (storyteller, teacher, enchanter), the third was the most essential: 'a great writer,' they learnt, 'is always a great enchanter'.[1] Although the term was firmly embedded in Nabokov's critical vocabulary as a label of praise, it is interesting that in his Harvard lectures on *Don Quixote* he also employed it pejoratively, cautioning his audience to remember that the Don's 'main enemies are enchanters'.[2] Nabokov did not greatly admire Cervantes's masterpiece – he dismissed it as 'a cruel and crude old book' in a 1966 interview (SO p. 103) – and the principal cause of his animus was the treatment meted out to the Knight of the Doleful Countenance by those seeking to 'cure' his delusions. Quixote's trusting imagination is exploited by the representatives of common sense in a spirit that evidently struck Nabokov as sadistically charitable, at best; and the defeat of Quixote's visionary gleam, signalled by his renunciation of knight-errantry, seems to have figured for Nabokov as a betrayal (principally by Cervantes) of that willing make-believe which is the primary impulse of literature itself.

1. Vladimir Nabokov, *Lectures on Literature*, ed. Fredson Bowers, (London: Weidenfeld and Nicolson, 1980), p. 5.
2. Vladimir Nabokov, *Lectures on Don Quixote*, ed. Fredson Bowers, (London: Weidenfeld and Nicolson, 1983), p. 93.

Leaving aside the question of whether Nabokov's was a just response to *Don Quixote*, we can see from the foregoing that he acknowledged two kinds of enchantment, good and bad. To illustrate this by the case of Don Quixote (whom in his comments Nabokov more than once paralleled with Christ): is his enacted dream of chivalry a distorting and imprisoning obsession, or a means of ennobling and transfiguring the world's cruel ordinariness? Nabokov would, I think, have judged Quixote as enjoying a liberating inventiveness of attitude rather than as suffering from manic delusions; but his own fiction shows how well aware he was of the vicinity of the two conditions, and of how it is not always easy to tell black magic from white. The novel in which this proximity is most memorably set forth is *Lolita* (1955); the two books with which this chapter is concerned, *Pnin* (1957) and *The Enchanter* (1939, tr. 1987),[3] contain separately elements that there are found combined. Both deal with the predicament of victims: Timofey Pnin is rather a Quixotic figure, persisting precariously in a world blind to the human value of his idiosyncrasies, which it views with sneering hilarity; and the anonymid who is the subject of Nabokov's 1939 novella is, like Humbert Humbert whom he foreshadows, the driven creature of his own perversion. Each of these shorter works reflects an aspect of the more complex *Lolita*.

Because of the legal difficulties that confronted *Lolita*, its publication in Britain and America was delayed until after that of *Pnin*. In fact, Nabokov had turned to the writing of *Pnin* (some of whose chapters were first published in the *New Yorker*) as he finished his labours with the *Lolita* manuscript; he wrote to Edmund Wilson that starting it felt like a sunny escape from the 'intolerable spell' of the longer novel. Originally envisaged as ten chapters, it finally extended to no more than seven, charting the vicissitudes in the life of its disconsolately resilient hero, with his endearing semi-assimilation to the America in which, like his creator, he has been transplanted. Like *Lolita*, it is both a tribute to America and – despite Nabokov's best intentions – a critique of it. In *Bend Sinister*, Krug's thumbnail sketch after a lecture tour of America had been of 'landscapes as yet unpolluted with conventional poetry, and life, that self-conscious stranger, being slapped on the back and told to relax'; but from the less distinguished Professor Pnin (pronounced P'neen), more effort is required: he is given room to survive, but not to put down roots. His rootlessness may be as much a factor of his selfhood as a consequence of uncertainties of

3. Vladimir Nabokov, *The Enchanter*, tr. Dmitri Nabokov, (London: Picador, 1987).

employment: the novel's curiously old-fashioned opening description depicts his 'strong-man torso' on its 'pair of spindly legs', as if to emphasize the slenderness of his connection to American soil. Happily adapting to national habits of dress, Pnin is unable to adapt to American habits of speech; out of time and out of place, he is freighted with a past that is both irrelevant to and inconceivable by the diminishing number of students who enrol for his classes. He has no child, he has no family, and virtually constitutes a one-man endangered species.

Nabokov's first novel, *Mary* (1926, tr. 1970), opened with characters in a lift stuck between floors in an émigré lodging-house in Berlin: an appropriate image for the arrested transitionality of those evicted from their past, but unable yet to form a future. *Pnin*, examining a different phase of Russian émigré experience, opens with our hero blissfully unaware of being on the wrong train, as the fate-figure of the conductor moves toward him down the aisle; but whilst offering us the chance to conceive him as a man on a journey to nowhere, this turns out to be the first of a series of small-scale disasters which are averted: in spite of missing the bus which would have corrected his error, Pnin nevertheless reaches his destination in time for dinner, and moreover does not, as feared, mislay his notes for the lecture he is to deliver there. Other reprieves granted during the course of the narrative are the heart-attack which looms but does not strike, his finding the way to a friend's place in the country when he had seemed hopelessly lost, the beautiful blue glass bowl which he fears he has broken but has not, and, finally, his triumphant surge toward future possibilities at the novel's end.

These are small but significant items on a balance sheet where for the most part red ink prevails. By the time the novel starts, in 1950, Pnin has lost his country, his language, and his wife; by the time it ends, five years later, he has lost his home (twice), his teeth, and his job. His tawny teeth can be replaced by sparkling dentures, but the home and the job are likely to be more problematical; the country and the language are irreplaceable. Like his creator Nabokov, Pnin left France for the USA in 1940; and since 1945 he has been teaching at Waindell College, a 'somewhat provincial institution' which may or may not owe something to the Cornell which Nabokov once described to Edmund Wilson as 'udder-conscious and udderly boring' (NWL p. 260). Pnin teaches Russian under the aegis of the German Department, whose head, Dr Herman Hagen, is his sponsor and protector; but who, in spite of attracting epithets such as 'good' and 'kind-hearted', is subtly exposed as suffering from that spiritual defect which, in Nabokov's fiction, is almost invariably the consequence of being German or of taking German culture seriously. Pnin's title of Assistant Professor – in which the grandeur of the second term is fatally modified by the first – marks him out as one of the vulnerable untenured; and his marginal, unfixed status is also expressed in the succession of temporary lodgings in which he briefly roosts. Born in 1898, he is rather old to be this insecure.

The campus setting enabled Nabokov to induge in (possibly therapeutic) satirical observations of American college life, something which he would repeat in *Pale Fire*. Waindell's buildings, with their 'ivied galleries' round an 'artificial lake' and their famous bells, are a pastiche of more dignified institutions; particularly rich in fatuity are its murals, depicting 'recognizable members of the faculty in the act of passing on the torch of knowledge from Aristotle, Shakespeare, and Pasteur to a lot of monstrously built farm boys and farm girls'. Fraudulent in its art and architecture, Waindell, led by its blind President, naturally prefers the specious or the vacuous as staff-members: a scheming careerist like Bodo von Falternfels (the cuckoo in Pnin's nest) or the rabidly jocose Jack Cockerell, head of English, are more representative appointments than scholarly Laurence Clements or a curio like Pnin himself. Most representative of all is the following: 'Two interesting characteristics distinguished Leonard Blorenge, Chairman of French Literature and Language; he disliked Literature and he had no French'; champion administrator, indefatigable conference-attender, and above all 'a highly esteemed money-getter', Blorenge more than compensates for any defects in his professional competence (he is also a personal friend of President Poore). Pnin's boat is finally sunk when Hagen, in a last-ditch attempt to find a haven for his obsolescent *protégé* approaches Blorenge; who refuses to employ Pnin in the French Department, once he discovers his ability both to speak and read that language. Nabokov had strong views about the way Russian language was taught at Cornell, which he eventually expressed in a letter (30 September 1958, L pp. 262–4); the situation was not without its parallels in *Pnin*.

Another target for scalding laughter is the Freudian excesses indulged in by Pnin's former wife, Liza. Like him a Russian émigrée, the two had met and wed in Paris in the late 1920s; after more than a decade of (vaguely visualized) marriage she left him without warning for a German, Dr Eric Wind, a 'totally humourless pedant' gifted with an understanding of her 'organic ego'. A year-and-a-half later, and seven months pregnant, she returned to Pnin in time for him to pay for her to accompany him to America; but alas, also aboard is Dr Wind, who punctures the bubble of Pnin's ecstasy by introducing himself during the voyage, explaining the necessity of their cruel ruse and – ever the gentleman – offering to reimburse Pnin half the cost of Liza's passage. They eventually marry (Wind had a wife to divorce), and embark on a mutual career in psychiatry, specializing in 'Group Psychotherapy Applied to Marriage Counselling' (the title of one of their joint-authored papers), in which groups of wives and husbands openly discuss intimate dissatisfactions and their possible source ('traumatic episodes floating out of everybody's childhood like corpses'). Eric Wind's ambition is to bring the groups of wives and husbands together in a still larger group; for he 'dreamed of a happy world consisting of Siamese centuplets, anatomically conjoined communities, whole nations built around a communicating liver'; Pnin, hearing of this, thinks that sorrow should be 'private': "Is sorrow not, one asks, the only thing in the world people really possess?". Liza is one of his own sorrows, to whom his loyalty is reminiscent of Cincinnatus's toward Marthe. She repeats her earlier abrupt re-entry to his life in 1952, and as before she wants something from him: preoccupied by the bad poetry of a new love-affair, she wants Pnin to keep a tutelary eye on her son Victor (the abdominal lump of 1940), whom she has enrolled in an exclusive private school nearby. Eric may have been the 'land father', she informs incredulous Pnin, but *he* is the boy's 'water father'.

As it happens, she is not so far wrong; Pnin would be altogether a more suitable father for Victor (the name nearly conferred on Nabokov by the muddled priest at his christening); and this in spite of the minor misunderstandings of their only meeting, where Pnin, expecting a small hungry boy who likes soccer and Jack London tales, meets a tall youth who hates sport, has no appetite, and has never heard of Jack London. His (un)natural parents had subjected him to a childhood structured *à la* Freud, but 'to the Winds, Victor was a problem child insofar as he refused to be one'; despite the 'psychometric' tests through which teutonically thorough Eric puts his son (at the hands of Louis and Christina Stern, whose clones turn up on Waindell campus as Christopher and Louise Starr), Victor uncooperatively produces 'drawings that had no subhuman significance whatever'. His gift as a painter, together with his IQ of 180, remain undiscovered by Eric and Liza, to whom the Sterns report that their son's 'artistic inclinations' entirely spoil the 'psychic value' of the tests he had undergone. In the gently eccentric Pnin, who at least knows something of the life of the mind, Victor might have discovered a father he need not have hidden from; for if in fact they confront each other with their differences, in fantasy they meet, and night finds them dreaming similar dreams (cf. pp. 71 and 91). This is not, however, a sentimental novel in which Pnin can compensate by adoption for the child he has never had;

just as Joyce in *Ulysses* hints that Bloom and Stephen could be for each other the father and son that each desires, only to acknowledge that this will not happen, so the relationship with Victor remains undeveloped – its potentiality, however, radiantly expressed in the water son's gift to Pnin of the (unbreakable) aquamarine punchbowl: 'What thou lovest well remains, the rest is dross'.

Nabokov despised Ezra Pound, but this line from Canto LXXXI well expresses how Pnin lives, how he survives. The book's main purpose is not to pour jovial scorn upon psychotherapy and academic mediocrity; its major theme is the beating (and intermittent spasms) of Pnin's loving heart, so many of whose fondest objects are in the inaccessible past. During the course of a lecture, Pnin is likely to indulge in a nostalgic ramble towards 'the days of his fervid and receptive youth (in a brilliant cosmos that seemed all the fresher for having been abolished by one blow of history)'; or he may observe, in semi-trance, long-dead relations and acquaintances in his audience: 'Murdered, forgotten, unrevenged, incorrupt, immortal, many old friends were scattered throughout the dim hall among more recent people'. Yet such passages remind us of just how severe the blows administered by history can be, and how acutely problematical such severance. Chapter Five is devoted to the visit Pnin makes in summer 1954 to the country house belonging to Al Cook (*né* Kukolnikov), where a reunion of Russian expatriates takes place biennially, enabling like-minded souls to share their impressions of the Soviet nightmare, or compare notes on the ignorance of their American students. In this congenial idyll Pnin seems in his element, discussing the chronology of *Anna Karenin*, swimming and sunbathing, transformed for the duration of a game of croquet from his maladjusted physicality to agile *victor ludorum*. If this be his element, however, it is bitter-sweet: here too, the past glimmers across its ditch of blood, and a casual conversation conjures up the suppressed ghost of Pnin's first love, Mira, who had escaped Lenin to fall prey to Hitler:

> In order to exist rationally, Pnin had taught himself, during the last ten years, never to remember Mira Belochkin – not because, in itself, the evocation of a youthful love affair, banal and brief, threatened his peace of mind (alas, recollections of his marriage to Liza were imperious enough to crowd out any former romance), but because, if one were quite sincere with oneself, no conscience, and hence no consciousness, could be expected to subsist in a world where such things as Mira's death were possible. One had to forget – because one could not live with the thought that this graceful, fragile, tender young woman with those eyes, that smile, those gardens and snows in the background, had been brought in a cattle car to an extermination camp and killed by an injection of phenol into the heart, into the

gentle heart one had heard beating under one's lips in the dusk of the past. (pp. 112–13)

After such knowledge, what forgiveness? This is a terrible item of mental luggage, marking its possessors out from fellow-travellers unscathed by history (even gentle Dr Hagen is distressed that the Nazis built Buchenwald so close to Goethe's Weimar); and this, of course, intensifies the introspective exile of such as Pnin, whose past contains these time-bombs, but who cannot for all that live in the simple present. Even the ambience at Cook's Castle, with its sunshine, its Russian cooking, and its shared reminiscences, cannot entirely overcome the awkwardness of being between two worlds, each equally inaccessible to those of a certain age:

> Some parents brought their offspring with them – healthy, tall, indolent, difficult American children of college age, with no sense of Nature, and no Russian, and no interest whatsoever in the niceties of their parents' backgrounds and pasts. They seemed to live at The Pines on a physical and mental plane entirely different from that of their parents: ... keeping always aloof (so that one felt one had engendered a brood of elves), and preferring any Onkwedo store product, any sort of canned goods to the marvellous Russian foods provided by the Kukolnikov household at long, loud dinners on the screened porch. (pp. 98–99)

As Pnin wanders beneath the trees, his heart gripped by the spasm of remembering Mira, he hears jazz from a car radio: the alien corn to which these young Americans are estrangingly addicted. The text exerts a two-way tolerance here: toward the forlorn impatience of the older generation, and toward the necessary self-centredness of the young; who have their own lives to get on with, and on an emblem of whose belief in the future (the silhouette of an embracing couple) this memorable chapter closes.

In the next, Pnin, 'battered and stunned by thirty-five years of home-lessness', believing himself on the verge of becoming a tenured member of the professariate, allows himself to contemplate the purchase of the small house he is happily renting. It falls to Dr Hagen to puncture this illusion. Pnin has declared to Victor that America sometimes surprises him, but always provokes his respect; his getting fired is one of its surprises. For the second time in the narrative (the third time in his life) his birthday signals his eviction from security; and Chapter Five's closing reference to 'Pnin's fading day' is clearer: for the future cannot be his – as it can be the embracing couple's. The final chapter sees the narrator, unobtrusively present throughout, take centre-stage as Pnin fades out – whose final, telephonically-transmitted words seem to announce complete obliteration: 'He is not at home, he has gone, he has quite gone'. What will remain at Waindell, it is clear, will be a set of colleagues locked into grotesque impersonations of the absent Pnin, as if in comment on their own essential nullity, too featureless to be burlesqued. Symbolically apt, the ending is not wholly sad: Pnin is not dead, for to be 'not at home' is the very essence of being Pnin; and he has 'gone', if not to a better place, than at least to a different place, while boobies like Blorenge remain where they are.

Pnin is usually considered the most approachable of Nabokov's novels, and the warmest-toned, in spite of the unhappinesses it documents; together with the portrait of the Swiss governess in *Speak, Memory*, Pnin stands as one of the most fully-achieved of his creator's characters, in spite of lacking the intellectuual distinction of a Krug or the diabolical charm of a Humbert. Yet, we might wonder, is there not some special pleading here? Why should Pnin, who is a rather unaccomplished, inefficient teacher and an unproductive researcher, be spared the narratorial spleen that vents itself so freely on Cockerell and Blorenge? The minor answer would be that, in an institution where so many mountebanks and charlatans draw salaries, a Pnin might at least be tolerated; the major question, however, does not relate to whether or not Pnin deserved tenure, but to why he deserved a novel: the answer to which is, because he is a human individual. In saying so, it is essential to understand that Pnin is not a type, a representative: we do not (like Melville's narrator at the end of his tale about 'Bartleby the Scrivener') exclaim 'Ah, Pnin! Ah, humanity!'; nor is he an everyman figure, like Willy Loman the dead salesman. Early on, the narrative had been at pains to discriminate Pnin from such stock samples as the absent-minded professor: for he is himself, an unique specimen on the verge of extinction, who appeals to both the artist and the scientist in Nabokov.

'Man exists only insofar as he is separate from his surroundings,' the novel asserts; 'it may be wonderful to mix with the landscape, but to do so is the end of the tender ego'. It is Pnin's separateness, the quiddity and tenderness of his ego, that the book celebrates; and it is all the more remarkable in that, unlike so many Nabokovian heroes, he is not a show-off or a monomaniac or a superb artificer. His truest note is perhaps sounded in the letter proposing marriage written to the trashy Liza, which the insensitive narrator copies out for us: 'I am not handsome, I am not interesting, I am not talented. I am not even rich. But Lise, I offer you everything I have, to the last blood corpuscle, to the last tear, everything. And, believe me, this is more than any genius can offer you because a genius needs to keep so much in store, and thus cannot offer you the whole of himself as I do'

(p. 153). Characteristically, the letter goes on to dispute the theory she espouses about 'birth being an act of suicide on the part of the infant': even while offering her everything, Pnin does not compromise his own beliefs. The fact that Liza was incapable of appreciating, still less of returning, such affections does not invalidate them – although it does lend a Quixotic lustre to their declaration.

Nabokov knew that people have hidden depths, that Etermons are individuals; he also knew that 'not even the enchanter himself' could be sure what the vessels in his conjuring set contained. But what about *Pnin*'s narrator, who shares so many of Nabokov's attributes, including his lepidoptery (rather fussily signposted): has he succeeded in pinning Pnin? Given Nabokov's intrusion upon Krug in *Bend Sinister*, we might equate the narratorial 'I' of *Pnin* with the author; but it is better to think of the narrator as a quasi-Nabokov, who at the last is eluded by the Pnin he has set out to describe. Due in part to the need for its chapters to stand as discrete episodes for the *New Yorker*, *Pnin* lacks both *Lolita*'s logodaedaly and its thematic complexity: the puzzles it sets are for the most part solved within the relevant chapters. There is nonetheless a continuous concern with design, and we note little markers of narratorial fore-knowingness; on one level, these imply the questions answered in the final chapter, about how the narrator knows so much about Pnin (he had been Liza's lover in Paris before Pnin, they have many common acquaintances); on a deeper level, they imply the questions of why Pnin so profoundly dislikes the narrator (only partly explained by the Liza connection?), and how much he finally does know about Pnin. These can seem arid considerations, of a type too frequently met with in criticism of Nabokov; but they bear upon the story's living pulse.

This novel does not wear its art on its sleeve, but most readers notice the squirrel motif, whose first instance is the poker-worked screen old Pnin remembers near young Pnin's bed, when his seizure in the park at Whitchurch immerses him in memories of a childhood fever. On it, a squirrel was holding something, and the delirious child is tormented by his inability to make the object out; similarly, he desperately tries to establish the horizontal pattern of recurrence in his bedroom's swirling wallpaper: were he to break that devilish code, health would return. As he was writing *Pnin*, Nabokov was also working on an annotated translation of *Anna Karenin*, drawing on the lectures he gave on Tostoy at Wellesley and Cornell; Pnin's discussion of the differential time-scale within that novel, and of the means by which its action can be dated, are directly due to his researches. The squirrel's hidden object is reminiscent of the nightmare both Anna and Vronsky have (we have already noted that Victor and Pnin share a dream), of a dirty little old man bending over a sack in which something is concealed, muttering in French about flattening iron; as Nabokov's explication makes clear, this is a foreshadowing of Anna's suicide beneath the goods train: in a sense, what is hidden in the sack is *her*. Pnin's screen, with its old man hunched on a bench and its squirrel, foreshadows himself on the bench in Whitchurch, and the grey squirrel with its peach stone that he sees as he emerges from his swoon; but the mood is less ominous than that of Anna's dream and, since Pnin is still alive (as squirrels so energetically are), the screen did not foretell his end: rather like Pushkin's failure to unriddle the date of his death, alluded to by Pnin in Chapter Three.

An aspect of that childhood fever had been his conviction that he was the object of a concealed pattern, imposed from without: life depended on finding the key. Yet Pnin survives in his ability to evade patterning, and to live in the implied trust that in time life will solve its own problems. 'Genius is noncomformity,' the novel asserts; and although Pnin is no genius he is touched by the saving ability not to fit in: he is a digressive creature, in his life as in his lecturing and his scholarship, and digression is the impulse to depart from pre-ordination (as in the last paragraph of Chapter Four). His life spontaneously makes room for the demands of others, be they ex-wives, water sons, thirsty squirrels, or the dog he adopts (Nabokov commented approvingly on Bloom's kindness to animals in *Ulysses*). One of the refreshing aspects of this novel is its evocation of the contingent world beyond Pnin's consciousness, the gratefully-heard hum of separate lives, where daughters marry and divorce, husbands refuse to go on diets, and even an unillusioned pederast can make a worthy life as an inspiring art teacher.

Awareness of these aspects of the novel enables us to understand the hostility Pnin feels toward Vladimir Vladimirovich, the narrator who has been zeroing in on him as if he were a prized butterfly; a hostility independent and surely in excess of anything in their shared pasts. In retrospect, the whole story moves toward a meeting between biographer and biographee, between stalker and prey: but at the last, Pnin leaps up and is gone, leaving V.V. with an empty net. An unflamboyant soul, Pnin had nevertheless protested with extraordinary public vehemence to blind bemused President Poore, about the plan by his compatriot Komarov, a painter, to insert Pnin's likeness on one of Waindell's murals, as an expression of Russian solidarity provoked by his dismissal. Again, Pnin's outrage is due to more than his personal and political dislike for the muralist, and surely indicates his profound sense of the obscenity of such a state of stasis: he won't consent to immurality, he won't be fitted

in. The patronizing V.V., offering Pnin a job, a niche, a starring role
in his next novel, is not dissimilar to those well-meaning busybodies
who harried Don Quixote back to 'sanity': these are versions of
immobility corresponding to Komarov's mural, deadly to the protean
Pnin of the croquet lawn. But just as young Huck Finn at the end
of his adventures runs away from 'sivilization', from language, and
above all from deceitful Mark Twain who had tricked him into
involvement with the book; so old Pnin in his little car, squired by
the unwanted dog his kind heart has adopted, sweeps majestically,
ludicrously past the shouting narrator, toward whatever windmills
he has yet to conquer.

Vladimir Vladimirovich, by contrast, is stuck in Waindell, at the
future mercy of such demented anecdotalists as Jack Cockerell,
manically embarked upon an untrue story about Pnin in the
novel's final sentence. Thus the book ends with a lie, and this
propels us to a reconsideration of the truthfulness or otherwise of
the entire account: does the narrator really know as much about his
subject as he has led us to suppose? Quite apart from Pnin's odd
outburst which he reports, in which Pnin asserts that V.V. 'makes
up everything' and 'is a dreadful inventor', there is an interesting
problem with dates. From carefully planted internal evidence, we
are able to deduce in Chapter Three that Pnin's security of tenure
as the Clements's tenant comes to an end on Tuesday 15 February,
1953: the Gregorian equivalent of his Julian Russian birthday. In
the final chapter, Vladimir Vladimirovich arrives in Waindell one
day early for his inaugural lecture, which he tells us is due to
be delivered on Tuesday, 15 February – the morning of which
gives him his last glimpse of Pnin. Pnin learns of his doom from
Hagen in the Fall term of 1954, so it is presumably the following
term, Spring 1955, that his redundancy takes effect. February 15
cannot fall on a Tuesday in both 1953 and 1955 (assuming the
earlier date is correct, 1959 would be the next recurrence); so
the narrator has his days or dates wrong. It is possible that the
mistake was Nabokov's, consequent upon the chapters' coming
into existence as short stories; but he was meticulous in the
matter of dates, and in his lectures (on *Mansfield Park*, for
example) he had a close eye for authorial chronology. Given that
in *Pnin* he draws attention to the matter of accurate dating in
the case of *Anna Karenin*, it is more likely that the solecism is
intentional, designed to be hard evidence of Vladimir Vladimirovich's
unreliability; which doubt once planted, we begin to wonder how
trustworthy can have been his effortlessly omniscient reconstruction
of Pnin.

'True art,' declared Nabokov in 1969, deals not with the genus, and not even with the species, but with an aberrant individual of the species' (SO p. 155); Pnin is certainly an aberrant individual, so Vladimir Vladimirovich meets that requirement. At times, too, his account takes inspiration from its digressive subject, as in this magic-carpet rumination on the problems of novel writing: 'Technically speaking, the narrator's art of integrating telephone conversations still lags far behind that of rendering dialogues conducted from room to room, or from window to window across some narrow blue alley in an ancient town with water so precious, and the misery of donkeys, and rugs for sale, and minarets, and foreigners and melons, and the vibrant morning echoes' (p. 26). It is no accident that Pnin himself is on the unseen end of the telephone; with some unwillingness we return from that Arabian day, evoked as if in his honour, to Joan Clements's hallway in Waindell. It is fitting that V.V.'s last glimpse of Pnin should also comprise a 'Moorish house' and a 'Lombardy poplar', each hinting of their exile from a far-away clime, yet also imparting to Waindell, like Pnin, a tang of the exotic. But if our narrator can pay such graceful tributes, he can also without qualm reproduce a love-letter of Pnin's, intended only for Liza's lethal blue eyes: a stunning invasion of privacy. There is a sharp contrast between such behaviour (worthy of a psychoanalyst) and Pnin's old-world social decorum, with its no-holds-barred generosity (for his party he makes punch with Chateau d'Yquem!). When first we meet Pnin, he has in his Gladstone bag his Cremona lecture, a student's essay ('Dostoyevsky and Gestalt Psychology'), and a lecture on 'Don Quixote and Faust'. The last item may merely be a dig at two overrated masterpieces (as Nabokov deemed them); or it may suggest two different kinds of temperament, two different modes of being and knowing: Don Quixote with his addled, idealizing imagination and altruistic divagations, contrasted to driven Faust, who sold his soul to gain inhuman knowledge. Perhaps it illustrates the difference between conscience and consciousness: Vladimir Vladimirovich may be one who has yielded to a Faustian temptation; in consequence, where Pnin suffers from his over-extended heart, V.V.'s problem has been with his eye.

The 'double time-scale' in *Anna Karenin* is an example of the relativity of time, Pnin (and behind him Nabokov) argues; 'reality' is not duration, because a soap bubble is as 'real' as a mammoth's tooth. There is a kind of double time in *Pnin*, composed of the present and the past; both spasmodically inhabited by Pnin, who lives in a different sort of time from Victor, the young American. There is of course an implied contrast between the old and the new worlds, between age and youth, which we also find in *Lolita*; there too we encounter something of the American specificity, the rendering of cultural detail, that is a feature of *Pnin*. But in *Lolita* this is crossed with another, much darker element that is absent from *Pnin*, and which had led their author to feel the later novel an escape from the former's 'intolerable spell'; for *Pnin* is an enchanting book, whose hero is freely, idiosyncratically himself – even if he becomes an *idée fixe* for others. *Lolita*, however, contains black magic; and the darker element setting off its scintillations was first explored by Nabokov in *The Enchanter*.

This title was the one intended by his father, according to Dmitri's translation of the Russian novella written by Nabokov in 1939. Published with some *éclat* in 1987 as his 'long lost novel', it is clearer now that *The Enchanter* is a lesser work, whose chief fascination lies in its relation to the vastly more substantial achievement of *Lolita*. Considered in that light, however, it has much to show us of the problems that Nabokov had to overcome before *Lolita* could be written. The basis of its plot was the same foundation on which the later novel's more intricate edifice is constructed: set in unspecified France, the protagonist, a forty-year-old diamond dealer, nurses a perverted fixation on pre-adolescent girls, which in the course of events tips over from contemplation into action. He cultivates the acquaintance of the (as it turns out, terminally) ill widow, whose russet-haired daughter had sparked his lust when he saw her roller-skating in a local park. As a means of securing access to the girl, he courts and marries the invalid who, after the false dawn of an apparently successful operation, obligingly dies. He gathers the girl from the foster care of her mother's friend, into a hired car with a seaside destination, where the hotel they stop at has only a double bed for Monsieur and his daughter to sleep in . . . at bedtime, slavering over her sleeping nudity, he goes too far and wakes her up to the sight of his none-too-magical wand, at which she screams the house down. Panic-stricken, he dashes out of the hotel into the path of one of the regularly-passing lorries which – in a parodic allusion to the end of *Anna Karenin* – flattens him.

The most emphatic difference between this story and Nabokov's later masterpiece is the extraordinary abstractness of *The Enchanter*: none of its characters has a name, none of its locations is specific. Its texture is in consequence impoverished, and this impairs the power of its illusion, even though Nabokov seems to have been trying to achieve the generalized feel of a fairy tale – albeit an inverted one. This approach also mimics the perception of the protagonist, who, outside of the specialized sparkles and details of his diamond-cutting, overlooks much in the outside world; apart, that is, from what relates to his sexual obsession, for it is no accident that the object most intricately and intimately visualized is the girl's naked body: 'His eyes returned from everywhere else to converge on the same suedelike fissure, which somehow seemed to come alive under his prismatic stare'. Here is another man with serious eye-trouble, as is also clear from the distortions wrought in his imagination upon the body of the girl's mother, his 'monstrous bride' and 'cumbersome behemoth', whose elephantiastic dimensions threaten to doom their wedding night: 'he (little Gulliver) would be physically unable to tackle those broad bones, those multiple caverns, the bulky velvet, the formless anklebones, the repulsively listing conformation of her ponderous pelvis' (Humbert Humbert will inherit such ungentlemanly reflections, but will spread a genial camouflage of wit over this stark evidence of pathological maladjustment). In the event, such fears are groundless; but the description eerily conflates his act of intercourse with one of murder: 'after the fact, it was with astonishment that he discovered the corpse of the miraculously vanquished giantess and gazed at the moiré girdle that almost totally concealed her scar'.

In such ways the prose attempts to reproduce the distorting lens of this unlovable jeweller's insane objectivity, unable to imagine the humanity of those he practises upon. But the third-person narration, curiously enough, rather than aiding such glacial detachment, hinders it; because when it connivingly enacts his frigidly libidinous imaginings or his inhuman evocations, the reader is placed in an uncomfortable voyeurism of events robbed of their first-person compulsiveness by the interpositioning of an implied narrator – whose dispassionate stance is problematical, because he has not (presumably) the excuse of clinical derangement. Our hero may have his twisted reasons for dwelling on the sexual details of a little girl's nakedness, but neither narrator nor reader has. Although Nabokov wished to distance himself as author from the sick soul he created, he was to understand by the time he wrote *Lolita* that such matters had to be the subject of a first-person account, a justifying 'confession' laid before judge and jury by semi-repentant, semi-elegant Humbert. If the fairy tale is to be inverted, it must be from the wolf's point of view.

Both *Pnin* and *The Enchanter* deal with displaced persons, but Pnin's exile is entirely different from the appalling estrangement of this distanced pervert; Pnin can (in Shakespeare's image) 'see feelingly', whereas the other is a coldly erotomanic eyeball. Insofar as he is an 'enchanter', his is definitely black magic: the sorcery that conjures up the beast within himself; the attempt to exert power over others, inanimating them in an inertly private fantasy (as opposed to the white magic which does not fear spontaneity, and empowers others through self-discovery). It has, however, to be questioned how spell-binding he is, in his attempts to turn his 'fairyland' into fact: he can practise upon the emotional vulnerability of a sickly, narcissistic woman (this cynical manoeuvering is the more despicable of his exploits), but he cannot compel her daughter's co-operation; when she glimpses his 'enchanted yardstick' all his power evaporates in the extremity of her disgust. In this bedroom scene, the title of 'enchanter' is ironically applied; he may possess a watch apparently without hands, figuring his hope to be exempt from time, but his middle-age can have no congress with her childhood, and he is exposed in all his bestial fraudulence. In fact, it becomes clear that, far from being a figure of power, he is himself helplessly at the mercy of his fantasies, so that when, in a singularly unpleasant image, 'an intolerable sensation of sanguine, dermal, multivascular communion' with the girl is evoked, 'as if the monstrous bisector pumping all the juices from the depths of his being extended into her, . . . as if this girl were growing out of him', this evolves into a state of umbilical dependence on *his* part. He is a type met elsewhere in Nabokov's fiction (Hermann in *Despair* is a case), of the lunatic schemer who imagines that the world conforms to his imposed scheme, only to find that no one else has been attending to the monotone, hypnotic whine of his defective selfhood.

Both Pnin and the 'Enchanter' could be described as non-conformist, but Pnin, with his ability to talk intelligently to a child, could hardly be more different from the latter, for whom a child is interesting insofar as it affords the opportunity for obscene gratification. Pnin's life has consisted of a series of diversions, exemplifing the valuable mode of marginality – like those quirky, interesting creatures which flourish in the borders of some illuminated manuscripts, spontaneously engendered by the playfulness of mediaeval scribes: bringing a sudden human element into a set text, and in certain instances encroaching so far that they overpower the central script. The 'Enchanter', by contrast, careers down the single track of his obsession to a predictable full-stop. Two types of imagination are implied by these two very different novels; but in *The Enchanter*, Nabokov has not in the method of his story fully matched the illicit daring of its theme: he did not attempt a sympathetic portrayal of the protagonist, and we all know that we dislike a pervert. In *Lolita*, his finest novel, the obsessive temperament and the digressive instinct, the despotic imagination and the life which struggles free, are brought together in a book which is a far more challenging example of compulsion and playfulness: the magics of art.

4

Dreamers and Demons

According to Humbert Humbert, he completed the manuscript of *Lolita* in 56 days; his creator Nabokov needed a good deal longer – nearly five years – to bring him and his confession into being. He would later consider *Lolita* to be his most difficult book (from the writer's, not the reader's, point of view), and it was doubtless 'the fascination of what's difficult', in Yeats's phrase, that made Nabokov turn again to the topic of *The Enchanter* a decade or so after setting aside his Russian version: to embark on the composition of what would be the most notorious, most famous, and least Russian of his novels. It proved no easier to publish than to write; in July 1954 he wrote in some despondency to Edmund Wilson of the two publishers who had already rejected it; he knew its themes were sensuous, but he insisted that 'its art is pure and its fun riotous' (NWL p. 285). Fear of prosecution for obscenity made both American and British publishers unwilling to take risks, and the novel was first published in Paris (1955) by the Olympia Press, which had a reputation for raciness. The debate about whether or not it was a 'dirty' book did *Lolita* no harm, and once it became clear that it would probably not be banned, more substantial publishing houses took an interest in it; Nabokov then encountered difficulties in arranging terms for a transfer of rights from the Olympia Press, so it was not until 1958 (US) and 1959 (UK) that the book was readily available, and flew to the top of the best seller lists. The novel which he had at one point thought of burning (his wife dissuaded him) remade Nabokov's fortune.

In the afterword written for the post-Olympian editions, he judged *Lolita* to have infringed one of three basic taboos of American publishing (the other two being the depiction of a happily pro-ductive black-white marriage, or the fruitful, contented longevity of a complete atheist). In the middle of what Robert Lowell has called the 'tranquillised' Fifties, an account of the sexual relations between a middle-aged man and a pre-adolescent orphan was not a promising subject for an author to propose to respectable publishers, on either side of the Atlantic. More than thirty years later, increased awareness of the incidence of 'child abuse' may give us, too, a hostile perspective: but the breaking of taboos and the extension of frontiers is the concern of the novel, far beyond its focus on the sexual specifics of the situation it sets forth: at one level of interpretation, the Humbert-Lolita relationship is a metaphor for other kinds of daring, transgression, and retribution. To make credible the *mésalliance* between sophisticated, intellectual Humbert and the much younger nymphet addicted to junk culture posed an obvious challenge to Nabokov's skill as a novelist; and the gulf between their two worlds may be an echo, albeit in a different register, of the disparities of experience, language, and learning that existed between Nabokov the teacher and the young Americans he addressed from behind his impenetrable lectern. The essential distinction was, of course, that Humbert resolved such differences by dominance in the bedroom; Professor Nabokov, by dominance in the classroom.

Every work of fiction must to some degree displace the real by the unreal in its reader's mind, commanding the willing suspension of disbelief, in Coleridge's classic formulation. *Lolita* takes to an extreme the task of making probable the improbable without resorting to any strategies of 'let's pretend', and of urbanely domesticating the reader's imagination amid a set of monstrous circumstances; and its outrageous matter contrasts with its relatively conventional manner. The book is, as has been pointed out, 'deliberately commonplace in romance structure',[1] with its elements of quest, attainment, journey, loss, pursuit, and revenge; not for *Lolita* the complications even of the triangular love-plot: here the pattern is a straight line, in accordance with whose unrelenting extension Charlotte Haze loves Humbert, who loves Lolita, who loves Quilty, who seems to love no one at all. What gives these relatively straightforward features their aberrant aspect is the extraordinary nature of the central relationship, and the exuberant intermediation of Humbert Humbert, the man whose wildest dreams come almost true.

Humbert appears a more successful magician than the furtive jeweller who was his forerunner; his zestful presence as narrator imparts a characteristic *brio* to his account – which, he tells us, he has for the

---

1. By Michael Long in *Marvell, Nabokov: Childhood and Arcadia,* (Oxford: University Press, 1984), p. 138.

most part kept uncoloured by the remorse which retrospection actually provokes in him. He tells us of his father, a Swiss hotelier in France, of the death by lightning of his mother when he was three, of his first nearly-but-not-quite consummated love-affair with Annabel Leigh at age thirteen. We learn of his studies, of his 'striking if somewhat brutal good looks' (repeatedly emphasized), of his fixation on 'nymphets' and the rather gruesome sex-life he has endured before his first marriage to Polish Valeria, in France, 1935 (the year of Lolita's birth); we learn of the end of that marriage in 1939 (the year Lolita's two-year-old brother died), on her refusal to accompany him to the USA, preferring to him a White Russian taxi-driver. Resident in America because an uncle made that a condition of bequeathing him an interest in his perfume business (Humbert's money comes from disguising animal odours), he tells us of his intensive researches toward a survey of French literature, of his trip to the Canadian Arctic, and of his three nervous breakdowns and consequent hospitalizations, after the third of which recuperating Humbert makes his way to Ramsdale, New England, where the family's house in which he had intended to spend his time in study and concupiscence (they have a girl of twelve) has burnt down, providentially or demonically diverting the lascivious lodger to Charlotte Haze's white-frame home. There, in late May 1947, Humbert encounters Lolita, a quarter-century his junior, and it is a case of loves at first sights: the widow's for the lodger, and the lodger's for her daughter.

The fate (or 'McFate' as jocularly personified by Humbert, after a classmate of Lolita's) that had till then conspired to interrupt or render ludicrous his transports, seems to shift into co-operative overdrive. Lolita has a crush on Humbert, by whom her mother's initially threatening proposal of marriage is soon seen as a chance to ensure easier access to the child. True, there are hiccups: Charlotte dislikes her daughter, perceiving in her a disruption of her own new-found bliss, and intends to marginalize Dolly as soon as possible; yet horror-struck Humbert finds he cannot dominate his second wife as easily as he had (or thought he had) the first. Her discovery of the journal in which he had incautiously recorded his infatuation with Lolita and his disgusted contempt for herself threatens an absolute rupture; but McFate steps in, killing Charlotte under a car on her way to post the fatal letters, which Humbert salvages; and, his marriage of convenience having inconvenienced him for only a few weeks, stepfatherly Humbert can pick orphaned Lolita up from the summer camp near Lake Climax to which her mother had banished her, and drive with her to an hotel at Briceland called 'The Enchanted Hunters' and, in mid-August 1947, at 6.15 one morning, can engage in the first of his copulations with the nymphet.

This takes us nearly to the end of Part One of this bipartite novel, with Humbert in triumphant possession of the hand (so to speak) if not the heart of his beloved, whom in its final chapter he showers with gifts:

> In the gay town of Lepingville I bought her four books of comics, a box of candy, a box of sanitary pads, two cokes, a manicure set, a travel clock with a luminous dial, a ring with a real topaz, a tennis racket, roller skates with white high shoes, field glasses, a portable radio set, chewing gum, a transparent raincoat, sunglasses, some more garments – swooners, shorts, all kinds of summer frocks. (p. 141)

Yet there's a certain accumulative desperation here, as the treadmill of his munificence turns arduously round; and how far short these everyday items fall – as did the suitcaseful of clothes with which he charmed her earlier – of the jewels cascading from an open casket, which alone would be appropriate to the noble note Humbert aspires to sound! These closing chapters show him at the height of his power, when for a transitory enchanted moment the circumstances of reality seem to transform themselves to fit his peculiar dream; but even here an ineradicable tawdriness is felt; his gifts are well on the way to being bribes, which in their turn will modulate into broken promises, then threats. It is certainly true that Humbert the perfumer has many an elegant locution to exalt his baser operations, beautifying the 'honey' of his spasms or bejewelling the 'hot, opalescent, thick tears' he sheds in their course; but in the end the reader, and Humbert himself, comes to suspect that it is all a case of floral decorations for bananas (to lift another phrase from Wallace Stevens).

In Chapter 14, having used Lolita as an unsuspecting aid to mastur-bation, Humbert congratulates himself on his success as a 'conjuror'. Musing on the event, he conceives that he made love to 'my own crea-tion, another, fanciful Lolita'; as he put it at the time, 'Lolita had been safely solipsized'. There is an element of strenuous self-justification here (did she really notice so little? or did *he* not notice her noticing?), but also a broader application; in that Humbert naturally tends to 'solipsize' the people around him, conforming them to his conceptions. Such transformative powers as he possesses, however, are a matter of his own subjectivity (and its verbal expression), rather than of any change wrought on the object; and even in the imperial security of his own prose, we can detect discrepancies between Humbert's version and the account that other participants might have given. His vigorous and distorting imagination hovers perpetually on the brink of delusion – as in the image of a distant denuded nymphet which, pursued, resolves itself into a man in his vest reading a newspaper.

Such episodes typify the rebuffs continually offered to the ardency of his imagination, whether by these bathetic revelations, or by the interruptions of his fine frenzies (the French widow – 'insolent hag' – who joins him on his park-bench where he's all a-tremble amidst nymphets, to ask if he has stomach-ache!). They lead toward the greatest discrepancy of all, between his vision of Lolita and the girl herself, who begins to destroy his dreams at the very moment she appears most to confirm them. Her refusal to be solipsized is prefigured by Valeria, the wife Humbert tried to content himself with in Paris. At first, she can do a passable imitation of a girl-child ('I derived some fun from that nuptial night and had the idiot in hysterics by sunrise'), but soon she is the target of the ribald disenchantment Humbert is apt to project upon any full-size female. Yet in spite of what he supposes to be her abject brainlessness, Valeria has wit enough to outwit him, acquiring a lover in whose taxi Humbert finds himself riding (a prefiguration of Quilty's car). Instead of his comedy wife – 'preening herself, between him and me, rouging her pursed lips, tripling her chin to pick at her blouse-bosom and so forth' – Humbert finds himself to be the comedy husband, prevented from beating her up by her stolid lover's omnipresence in their flat, exasperatingly genteel: 'turning away tactfully when Valechka took down with a flourish her pink panties from the clothesline above the tub' (that flourish is a brilliantly animating detail). The only retribution he can exact is to slam the door behind them – like a dentist's nurse in New England, thirteen years later.

As with Don Quixote, there is both cruelty and comedy in the repeated sallyings-forth of Humbert's imagination, encountering a world which obstinately won't live up (or down) to it; he does not, however, present himself as a victim (except in his more mawkish moments), and the cruelty tends to be projected outward by him, usually upon those who interfere with his designing. Chief amongst such obstacles, of course, is Charlotte Haze, whose ordinary female dimensions Humbert unchivalrously subjects to monstrous imaginative distention, just as he mordantly mocks her taste in interior decor, her affected French, and – most ignobly of all – her love for himself. There is little evident remorse when mischance reduces her to 'a dead woman, the top of her head a porridge of bone, brains, bronze hair and blood', in logical conclusion of the process of objectification he had already embarked upon. His tears do fall after the hilarious interview with Beale, driver of the fatal car who offers, 'with an air of perfect *savoir vivre* and gentlemanly generosity', to pay for

Charlotte's funeral ('with a drunken sob of gratitude I accepted it. This took him aback. Slowly, incredulously, he repeated what he had said. I thanked him again, even more profusely than before'); but these could be tears of relief. Yet helpless as Charlotte is in the web of Humbert's narration, we are free to wonder how much in her he missed or misunderstood, even though his focus is on how little she understood him; and we can derive some pathos from the fragments of the letters he has clawed to shreds and cannot then reconstitute, in which she stammers out her hopeless love.

Humbert has some resemblance to Hermann in *Despair*, both in his literary gifts (Humbert's are greater), and in his monomaniacal pursuit of his own plot and pattern, which blinds him to the plots and patterns of others. He can enumerate the hairs on a nymphet's forearm, but there are things which pass him by – her feelings, for example, and her interest in a man in a cigarette advertisement. When Charlotte shows him round the Haze home, Humbert, so full of seigneurial disdain for her folk-art ornaments and her toilet-seat cover, omits to interpret the tennis-ball, the browning (and Edenic) apple core and the discarded sock, that are the early warnings of Lolita's presence (the planting of these details by Nabokov is masterly). Early and late, she evades his fixative imagination: when, in Chapter 11, he becomes Humbert the Spider, spinning his sticky filaments throughout the house to try and catch Lolita, he has morosely concluded that she can't be inside when, at his very door, she chuckles that she has eaten his breakfast. For his part, Humbert would very much like to devour her, and (at a later stage) regrets that he can't 'apply voracious lips to her young matrix, her unknown heart, her nacreous liver, the sea-grapes of her lungs'; but within this verbal overlay there is 'a wincing child', as he can hardly bear to acknowledge, whose innards are more reliably read by the teacher who notes that she sighs a lot in class. 'Her unknown heart' would have been a quite appropriate title for Humbert's memoir; existing, like the other characters, in the refracting splash and swirl of his sea-changing prose, Lolita is almost a missing person: except when her sharp slang cuts through his display, or when in her surprising letter she simply declares, 'I have gone through much sadness and hardship'.

'I am not concerned with so-called "sex" at all,' blusters Humbert after their first night; 'a great endeavour lures me on: to fix once for all the perilous magic of nymphets'. He makes high-sounding claims on behalf of nympholepts such as himself, and of the 'incomparably more poignant bliss' they seek through union with a nymphet; to be a connoisseur in these matters is to be a poet rather than a pervert, he insists. A nymphet is marked out from her coevals by an undefinable something (not as conventional as good looks or intelligence, he asserts), by which she is identified as an inhabitant of an 'intangible island of entranced time'. It is a condition seldom encountered beyond the age of fourteen, and best discerned by a palpitating gentleman considerably older than the girl in question. In spite of Humbert's disclaimer, and whatever the transcendental emphasis with which he lards his explanation of these matters, it seems to be the case that 'so-called "sex"' is an indispensable part of the proceedings: an envisaged, if seldom attainable, end. Yet our imagination does not happily accommodate the penetration by a full-grown male of a girl merely on the threshold of puberty: there is a grossness, an unexpungeable impropriety; and the worst thing, for Humbert, is that he feels this too. High as his imagination soars, his 'tingling glands' bespeak earthbound appetencies. Keats acknowledged that 'The fancy cannot cheat so well / As she is fam'd to do, deceiving elf': as we have seen, the transformations wrought by Humbert's fancy are subjective, rather than objective; but by contrast, expressing his sexual urge replaces Humbert the man with Humbert the beast. 'The Enchanted Hunters' is named after its murals, which depict various huntsmen in a dreamy suspension of their blood-sport; Humbert snorts at the pretentiousness of this, but he has a different, unArcadian enchantment in view: the narcosis of Lolita into a 'completely anaesthetized little nude'. In *The Odyssey*, the enchantress Circe turned men into swine, and as Humbert and Lolita approached the fateful hotel a similar Circean transformation had touched the parked cars and the hotel receptionists; but this was a projection of Humbert's imminent swinishness.

Here again, his magic goes awry; the hotel itself is far from enchanting, his pills do not work, and it is an undrugged Lolita who, he tells us, initiates their sexual relationship. But far from feeling the release and ecstasy he was so sure would ensue, Humbert feels guilty, ill-at-ease; and indeed he has embarked on a fateful addiction, in the course of which the vocabulary he uses to describe their encounters declines from the floridly amusing periphrases to a deadened matter-of-factness: from honeyed spasms to 'a quick connection before dinner'. The lesson behind the episode when the glimpsed nymphet turns out to be a half-dressed male newspaper-reader is, that the vision depends on keeping one's distance: the greater Humbert's apparent intimacy with Lolita, the less she resembles the nymphet he craved. A parallel here is with Scott Fitzgerald's Gatsby, who at last discovers that Daisy is in actuality a far less enchanting and imaginatively compelling quantity than the green light at the end of her dock that had so long represented her for him. Just as Gatsby's dream of Daisy had 'gone beyond' the woman herself, so Humbert's dream had gone beyond Lolita, whose proximity in fact disables it: she is never less like a 'nymphet' than when they live together, whether in motel after motel or in their perilous cohabitation at Beardsley; never is he less the poet, more the pervert.

Part Two of *Lolita* recounts the disenchantment that results from the gross element in Humbert's nympholepsy overpowering its energies of the ideal. Much later, he understands how the delights as well as the torments of his condition had depended on his not attempting to make real his fantasies, thereby maintaining their visionary potency of 'the great rose-grey never-to-be had'. From the beginning there has been something self-defeating in Humbert's project to 'fix once for all' the allure of nymphets, because their chief attribute is an enchanted evanescence, inherently not to be 'fixed'. Having noted that 'the idea of time plays such a magic part in the matter', it is not surprising that Humbert is afflicted by a sense of the inevitable changes Lolita undergoes (he tabulates her growth), which will inexorably replace the nymphet he loves with the woman he fears. Like his forerunner the 'Enchanter' with his apparently timeless timepiece, Humbert hopes he might cheat time – 'never grow up' – but this madness is poor Humbert's enemy. It is significant that he mishears the name of the nearby lake as 'Our Glass Lake', transforming it in his imagination into a privately immobilized translucency, water miraculously stilled as glass; its proper name, of course, is Hourglass Lake – which doubtless denotes its shape, but also indicates that it too lies within the province of time: which Humbert underscores by swimming in it with his watch on (the 'waterproof' one given him by Charlotte: indestructible time!). The powerful car which panicky Humbert sees in his rear-view mirror has much in common with 'Time's wingèd chariot'.

In mythology, nymphs pursued with illicit amorous intent had a habit of transformation: Daphne turned into a laurel, Syrinx into a reed. The same is true of the nymphet; and Lolita's capacity for change is signalled, as early as the novel's second paragraph, by the several names by which she can be known: Lo, Lola, Dolores, Dolly, Lolita. The last is Humbert's intimate name for her, although he uses it less often to address her than its alternatives; and for her own part she does not use it – signing her letter to him 'Dolly'. On his first sight of her, he instantaneously linked her with his childhood sweetheart, Annabel: and it is as if he has succeeded, literally, in Proust's great effort to recapture lost time. He had been prevented from consummating his passion for Annabel, it will be remembered, by two ribald bathers emerging from the waves at precisely the wrong moment; sourly identified as 'the old man of the sea and his brother'. The Old Man of the Sea is another name for Proteus, the sea-god who keeps changing his shape (to avoid having to reveal the future); and it is indeed the constant intrusion of a world of change that prevents Humbert from fulfilling his fantasy of blissfully arrested time: not for nothing does he describe the relentless pursuer of Lolita and himself as 'a veritable Proteus of the highway'. When – after her disappearance from hospital in Elphinstone on 4 July 1949 – Humbert tries to trace this figure through the registers of the motels they'd stayed at, his unknown tormentor dissolves behind a series of mockingly cryptic aliases. Lolita, with her various names, is also a Protean quantity – which Humbert's chess-partner Gaston Godin seems dimly to have perceived, in asking how all his little girls are getting along.

The 'events' of Part Two consist of Humbert and Lolita's first journey through America (August 1947 to August 1948), their period of cohabitation at the college town of Beardsley, their second journey (dogged by the mysterious pursuer), Lolita's eventual disappearance and Humbert's frantic and futile attempts to find her; his liaison with Rita (who 'balances' Valeria in the structure), his reunion with Lolita (September 23, 1952), and his subsequent revenge on the man who had taken her from him more than three years earlier. So far from being a time of ecstasy, his two years living with Lolita are hellish in their insecurity, whether resulting from the thinness of motel walls or the inquisitiveness of neighbours in Beardsley. It is quite clear to Humbert, even at the time, that Lolita is made miserable by their parodically incestuous relationship: he hears her crying beside him at night, when he pretends to be asleep. With what he retrospectively sees as cruelty, he maintains his power over her by bribes, threats, and base stratagems (he routinely breaks promises, and steals the money she hoards toward eventual escape). The two of them exist in a state of mutual torment, his own sharpened by the knowledge that, whereas he cannot live without her, she could live without him, and wishes to do so. In Shakespeare's great sonnet about lust, the condition is defined as 'the expense of spirit in a waste of shame', in which the appetite, because directed toward an illusion, is finally insatiable. For Humbert, the 'joy proposed' of sexual intercourse with a nymphet turns, after the event, to nightmare; he finds himself 'more devastated than braced' by the satisfaction of his passion. Yet in its way this, like *Bend Sinister*, recounts the excruciation of an intense tenderness, with Humbert his own best tormentor as he recalls a dreadful intermixture of compulsions:

I recall certain moments, let us call them icebergs in paradise, when after having had my fill of her – after fabulous, insane exertions that left me limp and azure-barred – I would gather her in my arms with, at last, a mute moan of human tenderness . . . [which] would deepen to shame and despair, and I would lull and rock my lone light Lolita in my marble arms, and moan in her warm hair, and caress her at random and mutely ask her blessing, and at the peak of this human agonized selfless tenderness (with my soul actually hanging around her naked body and ready to repent), all at once, ironically, horribly, lust would swell again – and 'oh, *no*,' Lolita would say with a sigh to heaven, and the next moment the tenderness and the azure – all would be shattered. (pp. 284–5)

'None knows well,' Shakespeare's sonnet concludes, 'to shun the heav'n that leads men to this hell'; whatever its upward yearnings, Humbert's exasperated spirit is dragged downward to the sphere of his darker operations, and his lust intervenes between him and Lolita as effectively as does Clare Quilty. Quilty, their phantom pursuer who mockingly poses as Humbert's brother (*mon semblable! mon frère!*) to effect Lolita's discharge from hospital, is cast by apprehensive Humbert as a sexual predator: his car is glossily potent, its bonnet like a codpiece. When glimpsed at a swimming-pool, Humbert's feverishly distorting imagination – as with Charlotte – represents him as a blatant Priapus, 'his tight wet black bathing trunks bloated and bursting with vigour where his great fat bullybag was pulled and back like a padded shield over his reversed beasthood'. The Humbert who modelled swimming-trunks in front of a mirror has more in common with this figure than he may acknowledge; and, in turn, this image of rampant virility has little enough to do with Quilty, who turns out to be impotent, even if 'goatish'. In misrepresenting him thus, bipartite Humbert Humbert projects onto Quilty the unrestrainable male lust that is his own chief defect: Humbert the satyr figures Quilty as his capriform lower half, Hyde to his Jekyll; he bestializes him into a being antithetically lacking his own poetical capacities. To Humbert, Quilty is simple lust; whereas his own justification is that, in spite of everything, *he* loves Lolita. 'Love' is, of course, a tricksy term; as Blake's poem 'The Clod and the Pebble' illustrates, it has opposing aspects; for if the Clod sings of the selflessness of love ('Love seeketh not Itself to please'), its counterpart the Pebble, which has the last word, asserts:

> 'Love seeketh only Self to please,
> To bind another to Its delight,
> Joys in another's loss of ease,
> And builds a Hell in Heaven's despite.'

Such self-regarding manipulativeness certainly seems closer to Humbert's treatment of Lolita, up until the time of her escape to Quilty (significantly, on Independence Day). During this period, Humbert has little enough regard for her as an individual, although she is obviously necessary to him as a sexual adjunct. Indeed, just as he had initially planned to drug her, it continues to be the case that Humbert can only use Lolita by overlooking her as a person: her absence is required at their sex-sessions, and he condemns her selfishness in refusing to let him feel her up in the car whilst he watches other little girls going home. Considering the view he takes of Quilty, it is ironic that he owes him a considerable debt: for it is Quilty who, in luring Lolita away, breaks their imprisoning cycle of lust, and re-establishes the distance between Humbert and Lolita that enables him, not having her, to love her. Possessing her had made that love impossible; but once he has 'lost' her, Humbert's passion continues and acquires an aspect of altruism, so that at their final meeting, when she is so obviously no longer a 'nymphet', he realizes that he still loves her (although it could be argued, since here he notices her likeness to Botticelli's Venus, that he continues to substitute his vision of her for the girl she is).

Humbert's, however, is a case of love unreciprocated if not precisely unrequited: Emma Bovary, not Anna Karenin. Whatever the scope of his feelings for Lolita, she has virtually none for him – he is not even important enough for her to resent what he did to her. At their final meeting she makes it clear that Quilty has been and remains the love of her life. Humbert thus finds himself in the humiliating position in which Charlotte had been placed with regard to him; a humiliation sharpened by the awareness that Quilty, a minor playwright with a major appetite for pornographics and narcotics, is an altogether more vulgar specimen than himself. Of course, Lolita's passion for the ersatz glamour of such a creature tells us something about her; but it may also imply something about Humbert. Much as he despises Quilty, the latter was on Lolita's bedroom wall – indeed, had met Lolita – before Humbert; at the Enchanted Hunters, Humbert had parked his car in the space just vacated by Quilty's aztec-red convertible (Proteus would, of course, drive a 'convertible'; its 'aztec' colour alludes to Mexico, land of Lolita's conception and source of her name). Lolita's crush on Humbert originated in his resemblance to an actor whose photograph is on her wall with Humbert's initials affixed; but Quilty is there in his own right, advertising a brand of cigarettes. Had Humbert taken more interest in Lolita as a person, he would not have been deceived by her when she told him Quilty was female; for she had alluded to Quilty as a man, at the Enchanted Hunters. Humbert's overweening egocentricity prevents him from paying sufficient attention to others, until far too late; so it was with Valeria, so it is with Lolita: in the Elphinstone hospital his last gift to her had included Browning's dramatic works and a book on the Russian Ballet; it is only later he understands she would have preferred a comic, such as 'that repulsive strip with the big gagoon and his wife, a kiddoid gnomide'(!). When, after their final meeting, heart-broken Humbert described himself as 'drunk on the impossible past', his condition might well be diagnosed as the intoxication of belatedness – to borrow a phrase from Harold Bloom's criticism.

So central in his own life, Humbert is marginal elsewhere: it is significant that when in 1952 he looks up the *Briceland Gazette* for August 1947, hoping to find his image in the photograph he had intruded on, there is no trace of him (although there is in the text a trace of Quilty). Notwithstanding the vigour of his imagination and of his rhetoric, Humbert is curiously passive: his first wife divorces him, his second (to whose house he had come by accident rather than design) proposes to him; Lolita initiates their sexual relationship, then, in her turn, leaves him; and in spite of his concentrated efforts he is quite unable to find her or discover his rival's identity until she writes to him and then tells him about Quilty (with whom she had spent mere weeks in 1949, before being discarded). During the course of his account Humbert seems to dwindle, from the large-bodied male in congress with Lolita to the 'slender', 'fragile', 'diminutive' visitor to her shack on Hunter Road – a truly disenchanted hunter. By virtue of being its author Humbert wields great power in the tale he tells; but even so he finds himself conforming to the plans of others: clumsily playing his part – as he puts it – when Valeria leaves him and, having murdered Quilty, seeing it as the last act in the 'ingenious play' staged for him by that minor playwright – in which Humbert seems to have taken the role of second male lead. When shortly afterward he crashes his car, he anticipates with pleasure surrendering himself to the lifting hands of policemen and ambulancemen, deriving an 'eerie enjoyment' from his 'limpness'.

The plot to which Humbert must undeviatingly conform in his narration is of course the pattern established by what has already happened. Unable to recover time past, his revisitation of these scenes is – as he occasionally intimates – a journey along Memory Lane in which all his options have been foreclosed, and his imagination has in his own words been 'Procrusteanized' rather than 'Proustianized'. His story is a retrospection; its overshadowing sense is that of unalterably elapsed time, to whose stubborn structure he can do no more than add his own comic embellishments. These significantly serve to remind us this is more than a morality tale: there is much comedy when ass-headed Bottom becomes the erotic object of Titania, Queen of the Fairies, and this comedy also inheres in the Humbert-Lolita relationship; but Bottom did not – as Humbert does – sense his own bestiality; and Titania, unlike Lolita, *was* under a spell. The very incongruity of Humbert's attachment to Lolita offers insult not only to conventions governing sexual behaviour, but more essentially to time itself: born on 1 January, Lolita represents all things new, and through her he seeks to drink at the wellspring of youth; he succeeds merely in polluting its waters. In retrospect, Humbert comes to appreciate what he had violated in Lolita; and the account he offers is, he tells us, an attempted cure for the remorse he feels, 'the melancholy and very local palliative of articulate art'.

There is much death in the book: set-pieces like Charlotte's accident and Quilty's murder stand out, but there are side-shows of mortality – Jean Farlow's fatal cancer or Charlie Holmes's death in action, in Korea. Humbert's mother, first love Annabel, and first wife Valeria are items in a pattern of fatality that only Rita, of the women associated with him, escapes. All these deaths are premature, and Valeria's death in childbirth reminds us that there are dead children in *Lolita*, as well: Annabel, Lolita's younger brother, and her own stillborn daughter come to mind; Mona Dahl's mother has a baby that dies in infancy, and Humbert remembers or imagines motel notices forbidding guests to flush stillborn babies down the toilet. Lolita's own death in childbed, of which we learn in John Ray's foreword, takes place at the Alaskan antipodes of her Mexican conception on 25 December 1952; and this dead mother and unliving child seem to negate the message of hope traditionally associated with Christmas Day. John Ray also gives us the date of Humbert's fatal heart-attack (16 November 1952), which is meant to have occurred immediately after his completing the final paragraph (in a 1967 interview, Nabokov explained the remoteness of tone of this paragraph, with its abbreviated references to 'CQ' and 'HH', as intended to suggest Humbert's hurried response to his cardiac intimations of mortality). For once, however, Nabokov seems to have made a slip: if we count back from Humbert's death the 56 days it took him to write the manuscript in legal captivity, we arrive at 22 September – the date on which he received Lolita's letter (interestingly, in the previous chapter he talks of receiving it 'early in September'; in his Russian translation Nabokov corrected this to 'late in September'). He was not arrested until 25 September, which would have been the earliest possible date to start his confession.

Nabokov claimed that the initial impulse to write *Lolita* came after reading a newspaper report of an ape in a Parisian zoo which, given charcoal, paper and much encouragement, had eventually produced a crude drawing: the bars of its cage. So far, researchers have failed to trace it, but doubtless this very item was what the man-mistaken-for-nymphet was reading: for Humbert's story eloquently outlines the cage-bars of his nympholepsy. A similar subterranean contribution may have been made by an American short story, first published in 1858 but subsequently anthologized, called 'The Diamond Lens' by Fitz-James O'Brien (who died fighting for the Union in 1862). This tale owes some debts to Poe, who is of course evoked and alluded to in *Lolita*; and I wonder whether it, along with Poe's tales and those of writers like Jules Verne and H. G. Wells, might not have been encountered by the young Nabokov in his father's extensive library (or he might have come across it after 1940, as he acquainted himself more thoroughly with the literature of his adopted country). We are not talking about anything as straightforward as influence, still less of a source; rather, of a piece of writing which may have left a faint deposit in a retentive mind.[2]

'The Diamond Lens' is narrated by one who from his childhood on has been wholly fascinated by 'microscopic investigations'. Such is his passion that as a ten-year-old he had been made a present of a basic microscope; which enabled him to enter the invisible world – 'the dull veil of ordinary existence that hung across the world seemed suddenly to roll away, and to lay bare a land of enchantments'. He stood to inherit an aunt's wealth, but even so his parents insisted on his training for a profession, so under pretence of studying medicine he removed to New York; while in fact he continued to be entirely given over to microscopy, spending large sums on equipment which nonetheless falls short of his ideal. In his apartment block he has for an upstairs neighbour an exotic personality, Jules Simon, whom he supposes

2. It is of course only my hypothesis that Nabokov had read this tale (which I myself read as a boy, in an anthology). On rereading *Lolita* for this study, I found myself remembering the microscopic details of the story (although not, interestingly, the murder), and wondering about parallels with Nabokov's novel. Unable to remember either author or title, and unable to find anyone else who knew the tale, I eventually ran it to earth by means of an Index to short stories, under 'microscopy'. It is probably most easily available in *Future Perfect: American Science Fiction of the Nineteenth Century*, revised edition, ed. H. Bruce Franklin (New York: Oxford University Press, 1978), pp. 328–51. I was interested to find that Franklin alludes to Humbert Humbert in his introduction to the tale (p. 323), so the congruences between it and *Lolita* are not wholly imaginary. The means by which I located this tale confirmed for me Nabokov's sense that time past is recovered by conscious labour rather than by intuition.

to be a Jew, and from whom he has bought the occasional antique curio. Simon tells him of a remarkable spiritualist medium, whom the narrator visits (this is when we learn his name, Linley), in order to place himself in communication with the spirit of Leeuwenhoek, 'the great father of microscopics'. The séance is an astounding success: he is given a lucid answer, in which he learns that he needs an enormous diamond to make the most powerful lens possible. The problem of how to come by it is solved on his return, in a state of high excitement, when he discovers that Simon himself has such a gem in his possession, illicitly. Linley overcomes Simon's suspicion with drink, and then murders him; arranging matters to look like suicide. Successful in this deception, he creates the marvellous lens, by whose enormous magnification he is at last able to discern, within a droplet of water, a whole enchanted garden-glade, complete with a perfectly-formed, unutterably beautiful (and, we infer, totally nude) sylph-like female, whom he calls 'Animula'. He falls impossibly in love with her, and is of course tormented by the gulf nature has set between their orders of existence. His health suffers. Trying to drive out fire with fire, he betakes himself to the *risqué* cabaret performance of a 'celebrated *danseuse*', but this athletic giantess disgusts him, and he returns to his vision in the microscope; only to discover that there Animula is ailing, shrivelling before his horrified gaze: too late, he realizes that what is happening is that the droplet has all but evaporated. He beholds her death-throes, then in a frenzy destroys his instrument. He recounts his story from a present in which he is mocked as 'Linley, the mad microscopist'.

The tale is better than this bald summary suggests; it is as good as the best of Poe. It offers several intriguing congruences with *Lolita*. There are minor aspects, such as Linley's legacy from his aunt and Humbert's from his uncle; or the murder-scene with the drunken (and melodramatic) Simon, which has some affinities with Quilty's death (Linley hears 'a smothered sound issue from his throat, precisely like the bursting of a large air-bubble, sent up by a diver, when it reaches the surface'; Humbert sees a bubble of blood form on Quilty's mouth). Of greater significance is the description Linley gives of his passion for microscopy, whose gradual occlusion of other aspects of his life mirrors Humbert's absorption in his nympholepsy: it is, he says, like 'the pure enjoyment of a poet to whom a world of wonders has been disclosed', it is like discovering Eden, and resolving to 'enjoy it in solitude'; it makes him feel superior to his fellow-men ('I was in daily communication with living wonders, such as they never imagined in their wildest visions'). This is not unlike the claims Humbert makes for the nympholept, and in each case the conviction of superiority conduces to the moral ease with which a murder is committed. Just as McFate initially enters into apparent collaboration with Humbert's designs on Lolita, so for Linley everything he desires falls into place ('Here was an amazing coincidence. The hand of Destiny seemed in it' – on his discovering that Simon has the diamond he needs). Humbert's maladjusted vision resembles the microscopist's eye-view – not least in their disgust for full-sized women ('those heavy muscular limbs, those thick ankles, those cavernous eyes', fumes Linley at the dancer's *'pas-de-fascination'*). In each case, a destructive and impossible love ensues, between a man and a disproportionately small female, and the collusion of fate is suddenly and catastrophically withdrawn. Both offer first-person narrative accounts of their fatal fascinations, in an attempted repudiation of the anticipated judgement that the writer is a madman. Humbert has lost his nymph, Linley has lost his sylph (who is like a 'graceful naiad' moving in the 'unruffled waters that fill the chambers of the sea',[3] but who, like Lolita, displays an 'innocent coquetry'). Nabokov himself was no stranger to the excitements of microscopy, and if he read 'The Diamond Lens' as boy or as man, could hardly have failed to respond to that alone; but it is interesting that, as in *Lolita*, a different order of morality is claimed by O'Brien's protagonist, who is a man beyond: and just as Humbert feels remorse at his treatment of Lolita but none for Quilty, so Linley (whose story may, of course, be a Kinbotian fabrication to explain his failure in life) mourns the loss of Animula, but has no compunction over the murder of Jules Simon. Both murderers have fancy prose-styles, in which an imprisoning obsessiveness allied with a creatively distorting imagination hold a cracked mirror up for their readers.

There is much death in *Lolita*, but there is also much laughter: we noted earlier Nabokov's declaration to Edmund Wilson that it exemplified pure art and riotous fun. The foregoing discussion has risked over-emphasizing the moralistic aspects of his tale which, notwithstanding its dark elements, is extremely funny; and recognition of its humour is essential to understanding its achievement. The novel's comic aspects derive from the disparities with which it deals. 'When tender youth has wedded stooping age', declares Chaucer's Merchant, 'There is such mirth that it may not be written'; Humbert, not exactly stooping (and not exactly wedded), in writing the unwritable extorts considerable humour from the circumstances he depicts; and this is

---

3. One can of course take such matters too far; but I wonder if the young T. S. Eliot did not read this tale, and give this image and this phrase to J. Alfred Prufrock, as he remembers his lost mermaids.

rendered more extravagant by the contrast with his subject matter, which would seem to require a sombre, penitential tone. Critics who complained that Humbert was in no position to criticize the vulgarity he detected in others, missed the point that the disparity between his private conduct and his public dignity leads to amusement as well as disgust: there is a superb outrageousness in his hypocritical fastidiousness. It is this quality of knowingly irresponsible imagination that underlies many of his sallies; as when, for example, in his awkward and dangerous interview with Lolita's headmistress, he suddenly muses, 'Should I marry Pratt and strangle her?' (this is the occasion when he learns of the shockable Miss Redcock, soon to be married; and gives his permission for Lolita to participate in the school play 'provided male parts are taken by female parts'). Reflecting on the coy personifications in a girl's magazine ('Mr Uterus' making a thick wall to bed down a potential baby), Humbert sarcastically evokes a 'tiny madman in his padded cell'.

A similar sarcasm, whether scorching or morose, often informs his reaction to America, its motels and its teenagers, its market-led culture (not for nothing is Lolita once described as 'glad as an ad') and its schoolmarmish notions of respectability; a country where 'ART' stands for 'American Refrigerator Transit'. Unenthusiastically he lists their ports of call as they criss-crossed America, his wearied contempt best summed up by the commentless itemization of various local wonders on which his eye has lustrelessly dwelt; archetypical of which is the zoo 'where a large troop of monkeys lived on a concrete replica of Christopher Columbus' flagship'. From his later vantage-point he sees this decadent Odyssey as a pollution, reducing the 'lovely, trustful, dreamy, enormous country' to faded road-maps and tourist guides; and much of his disgust is, as we have seen, projected self-disgust. One piece alone of vulgar America withstands his sneering scrutiny, and that is Lolita herself: from her first words – 'The McCoo girl? Ginny McCoo? Oh, she's a fright. And mean. And lame. Nearly died of polio.' (her comment on the 'nymphet' Humbert had anticipated lodging with) – to her last ('Good by-aye!'), she does nothing to suggest that, in human terms, she is at all extraordinary. Indeed, as Humbert recognizes, she is in many respects extremely conventional; that he adores her in spite of this goes to show that 'love is blind all day and may not see' (Chaucer again); but also reminds us of the attractive zestfulness in the very vulgarity he affects to depise.

As unlikely as the relationship between Humbert and Lolita, is the accommodation made between his high style and the unbuttoned, slangy creativity that is her verbal element ('Yessir! The Joe-Roe marital enigma is making yaps flap'). Nabokov had enormous fun introducing the two to each other; he once said that the drawback of knowing three languages was trying to keep pace with their ever-changing slang; and slang, as the informal, rule-breaking use of language, whose best effects come from its explicit deviation from dictionary decorum, is intrinsically humorous: insofar as humour represents the free play of the spirit in the margins of necessity. In his afterword to *Lolita*, Nabokov alluded to his own private tragedy, the forcible linguistic expatriation from his Mother Russian; yet this, his best novel, is the one with the least of Russia and the greatest American specificity. Again and again Humbert's old-world locutions and pseudo-genteel mannerisms collide with Lolita's barbaric yawp (or Bronx cheer), as two worlds meet with a consequent jarring of hemispheres: as when Humbert in the Haze home hears Lolita yell to Charlotte that *she* has the magazine her mother had just asked him for; provoking the exasperated reflection that 'we are quite a lending library in this house, thunder of God'. The interplay between Monsieur Humbert's disdainful *hauteur* and the scruffy unselfconsciousness of the nymphet, by whom he is as mesmerized as he is repelled, is a richly comic resource.

But it is more than that, of course. Compulsion and playfulness, imprisonment and attempts towards freedom, intertwine at the heart of the novel. Lust-driven Humbert is imprisoned in his nympholepsy, with its terrible fixation on a particular – and almost invariably disappointing – goal; yet he is, as he tells us, highly susceptible to 'the magic of games', and in the novel this is best exemplified by Lolita's tennis: which is fluent and graceful, and at the same time innocent of all desire to win – in this, offering a contrast to his chess-playing. Humbert the dominator regards it as a fatal flaw in her game; yet in her elevation of the aesthetic over the competitive elements, in freeing herself from the will to achieve, Lolita seems truly to inhabit that higher plane, which he attempts to convince himself *he* can enter upon by having sex with her. His passion for her is to the lowest degree goal-oriented; whereas her tennis exemplifies that 'purposiveness without purpose' which Kant offered as the definition of art (we might compare Lolita's tennis strategy with the way Holden Caulfield's puppy love, Jane Gallagher, played checkers in *The Catcher in the Rye*: she would keep all her kings in the back row, because she liked the look of them there). Sexual love may permit the most ardent conjunction, not only between two people, but also between an individual's spiritual and physical parts (such a dualism can, of course, be disputed); but Humbert remains a centaur, a riven creature, because his preferred sexual partner cannot be a true partner in any other aspect of his life. In acknowledging this rift in him, however, we see that he is more than a pervert; and *Lolita* is more than a study of perversion – even though Nabokov did some basic research into such deviant types, in order to write it. Those who suffer from perverted appetites seem often to wish to represent themselves as to some degree 'normal', and their desires as more ordinary – because more widespread – than is customarily supposed (hence such sorry associations as the so-called Paedophiles Society). Humbert, on the other hand, wishes to proclaim his specialness: on the face of it, he argues that his sexual proclivity is one characteristic of his superior uniqueness.

Given that this is not a fictionalized case-study of a pervert's mind, and that Nabokov was not engaging in any crusade on behalf of child-molesters, what were the attractions of the subject for him? Its difficulty, as was noted at the outset, must have posed an intriguing challenge: how to write a book about an adult's sexual fascination for a child, that was not a predictable denunciation; how to beguile a reader by Humbert's ingratiating arrogance; how to portray a murderer with a fancy prose style; and how to make the whole vigorously comic, but at the same time plausible. To overcome the inherent improbability of the relationship between Humbert and Lolita, and to depict this fabulously dangerous liaison taking place in a convincingly-rendered small-town America, was for Nabokov to confront in an extreme form the fiction-writer's task of importing the unreal into the real; when exotic Humbert and hum-drum Lolita cohabit in the reader's imagination, it is a triumph of technique and of fictional excess. Chaucer suggested that the 'mirth' arising out of the union of such opposites could 'not be written', whether because it was too difficult or too indecent to attempt; but Wallace Stevens saw that the very processes of fiction imply an overthrowing of decorum, and that in dreams begin irresponsibilities: 'fictive things,' he asserts at the end of his poem to 'A High-toned Old Christian Woman', 'wink most when widows wince'. The reader-cum-juryman envisaged by Humbert as the recipient of this tale, whose eyebrows he supposes will have travelled halfway round his head with shock, begins as his adversary, but becomes his collaborator: we forget to be as appalled by the murder of Quilty as we were by his use of Lolita, and are inclined to fall in with Humbert's insane supposition that he had the right to exact revenge.

Nabokov, encountering difficulties in finding a publisher for *Lolita*, judged that this was because it infringed a publishers' taboo. He should not have been surprised, given that taboo-breaking and the extension of frontiers are so central to the novel's methods, as well as to its subject. When near the end Humbert deliberately insults his dentist (Nabokov had had painful encounters with the profession), or samples the exhilaration of driving on the wrong side of the road, these are simply more clear-cut instances of the inherent outrageousness that has marked, not only his behaviour, but the novel itself, presenting for our fictional delectation matters we would normally address with sombre earnestness. There is a liberating excessiveness in Humbert's zestful brutality toward the dailiness of life, which is allied to his creator's sense that art itself involves a breaking of boundaries and an expansion beyond ordinariness – seen, for a minor instance, in the fine surprise of the stiff-legged kangaroo bounces with which Humbert cuts off Quilty's escape-route in Pavor Manor, and for a major instance in the extraordinariness of the entire murder-scene (a set-piece which was one of the earlier parts of the novel to be composed). 'I have written a wicked book, but feel spotless as the lamb', wrote Melville to Hawthorne, having finished *Moby-Dick*; *Lolita* offers readers the affront of evoking our laughter whilst flouting some of our pieties; and it achieves this through the purity of its art, which applies a high style to a low life.

Humbert informs us that in writing his account he has permitted the artist to take precedence over the gentleman (just as in life he had let the lecher do so). Early in the composition Nabokov described him, in a letter, as a 'very moral middle-aged gentleman' who finds himself 'very immorally' in love (12.11.51 L p. 128); and reviewers wishing to defend the novel from accusations of obscenity stressed how moral was its message. This position had, however, been foreseen and parodied at the end of John Ray's foreword, where he adjudges the book's 'ethical impact' to be of greater significance than its 'literary worth'; thus flatly contradicting the profoundest belief of his creator, for whom the value of *Lolita* was in the inherent morality of its uninhibited art. The book's worth lies in its uninhibitedness rather than in any affirmation of well-regulated conduct, and *Lolita* is most itself in its forays beyond the gentility principle – just as Humbert is most truly himself in the singularity of his imagination. The distinctive and challenging quality of the tale he tells is that its shocking subject-matter is presented wittily and with enormous sophistication; few things could be further from the sordid actualities of his virtual raping of a minor, than the inventive brilliance of his narration: and this disparity between the elevation of the style and the baseness of the theme goes to the heart of this remarkable book, which in both matter and manner amounts to a repudiation of decorum.

Sex, Humbert asserts, is the ancilla of art: its servant, not its source. In *Lolita* art and sex are simultaneously like and unalike: like, in that Humbert's sexual nature deviates from and protests against the conventional, just as art, imposing form, implicitly protests against the formlessness of mere being: in Henry James's terms, art's 'sublime economy' is applied to the 'splendid waste' of life. Both as poet and as pervert, Humbert is an exceptional man. The dissimilarity is that his art is the means by which he achieves freedom from the imprisonment of time and of his sexual appetite. From his point of view, the writing of the tale becomes an act of expiation whereby he atones in retrospect for sins committed unrepentantly, and transmutes his selfish lust into a selfless love: 'articulate art' is a 'palliative' for his condition. From the reader's point of view, the novel is a demonstration of an exaltation of artistry that transcends what James termed the 'fatal futility of fact': the ugliness of Humbert is finally transfigured, his ass's head replaced by (super)human features, as 'clumsy Life' (Henry James again) turns into something rich and strange. This transformation is both magical and outrageous, since Humbert the artificer supplanting Humbert the pervert is only partially penitent.

In his case, art is to redeem what sex destroyed; through his perversion he despoiled Lolita, through his art he immortalizes her – and himself. In spite of what he implies to be his finally selfless devotion to Lolita, she is his ticket to eternity, and their story, which ends evoking their potential immortality in art, depends on death: Humbert's, simultaneous with the final full-stop, and more importantly Lolita's, since he makes it a condition of publication that she be dead. He had supposed she would live into old age, but this, like so much else about her (including the sex of her unborn child), he gets wrong. After their final meeting he describes her as both 'dead' and 'immortal', hurriedly glossing this by explaining the terms governing publication of his manuscript; but it seems no less true to assert that she can only receive immortality at his hands by dying to the world. He is, in truth, a reversed Pygmalion: that sculptor created so perfectly beautiful a statue that he fell in love with it, and by his love brought her to life; Humbert encounters a girl, for whose living actuality he substitutes a perfectly-imagined creature whom he loves; and to bring Lolita to birth in art has required her death in life. The 'prophetic sonnets' to which the last paragraph refers are presumably Shakespeare's, in which the poet repeats to his beloved that whereas s/he must in time die, to be the subject of his verse ensures eternal life. As the old saw has it, *ars longa: vita brevis.*

So too Keats, contemplating the immobile immortality of the figures on his Grecian urn, reassures the 'bold Lover' he sees there that his beloved 'cannot fade', and that 'for ever wilt thou love, and she be fair!'. Yet as the poem suggests, this suspended animation has a cold air about it. Humbert, finally assimilating his Lolita to Botticelli's Venus – despite the resistance she offers of being dowdy and pregnant – may have purged himself of his baser elements (symbolically slaughtered with Quilty); but he confers on her the gift of a dubious immortality, not unlike the Duke in Browning's 'My Last Duchess', who replaces the wife he has done away with by a breath-taking portrait of her 'as if alive'. Like that Duke's durable pigments – and like Bluebeard's castle – the 'refuge of art' Humbert offers his Lolita is an ominous sanctuary, in which she finds herself safely solipsized. All along he has loved a Lolita who was not there, and the culmination is to replace the original of Lolita by a creature commensurate with his capacity for wonder, an opportunity for the consuming voracity of his imagination. Love is by its nature powerfully idealizing, but the beloved should not become its almost incidental object; this way madness lies, and that is why Theseus, in *A Midsummer Night's Dream*, associates the 'lover' with the 'lunatic' as well as with the 'poet'. *Lolita* turns out to be a monument to Humbert's triumphant effacement of Lolita herself, who we might say has been canonized for art. This may well exemplify what Henry James called 'the madness of art'; for Humbert's fiction, like his nympholepsy, enacts a struggle with time, and the extravagances of his imagination (that 'supreme delight of the immortal and immature', in Nabokov's words) may – again like it – have a pathological aspect. When, as readers, we find ourselves compelled by Humbert's fancy prose-style even as we deplore his morals, it is difficult to be sure whether dreamer or demon has the final word. We can at least be certain that this is a novel in which fictive things wink at us fast and furiously.

5

Readers and Writers

Like *Lolita*, the germinal beginnings of *Pale Fire* can be traced back to Nabokov's last months in Paris at the outbreak of war, when he projected a novel to be written in Russian and titled *Solus Rex*. All that remains of this is two stories, intended as its first and second chapters, published in 1940 and 1942 and subsequently translated as 'Ultima Thule' and 'Solus Rex'.[1] Eighteen years later, having finished *Pnin*, and as he brought to a close his years of labour on the translation of, and monumental commentary on, Pushkin's verse-novel *Eugene Onegin*, Nabokov turned again to some of the themes of *Solus Rex*. In a letter of 24 March 1957 to a prospective publisher, he gave a rough outline of his proposed novel, by then called *Pale Fire*: it was to concern the monarch of 'Ultima Thule, an insular kingdom', who has been forced into American exile by a palace coup abetted by neighbouring 'Nova Zembla'; the King was to be pursued by a secret agent, Mr Copinsay (of Orcadian descent), who makes a bungling progress toward his intended victim – living incognito 'with the lady he loves'. Assassin and reader would be surprised by what happened, declared Nabokov; and he also projected an American setting whose basic realism would be tempered by certain magical readjustments. These notwithstanding, 'from the picture window of my creature's house one can see the bright mud of a private road and a leafless tree all at once ablaze with a dozen waxwings' (L pp. 212–13). On 8 April that year Nabokov signed a contract with G. P. Putnam's, and accepted an advance of $2,500 on *Pale Fire*.

Both letter and contract suggest confidence; but, again like *Lolita*, this was not an easy book to bring forth, for in August 1959 he wrote

1. These stories are described in Boyd (pp. 517–20).

to Putnam's, asking to be released from his contract and offering to repay the advance. The work had not been progressing well, and he had reached the conclusion that his very 'contractual obligation' to produce the novel was an interference with its 'free development'; 'I am not sure I shall ever write it,' he said, 'but if I do it will not be in the near future' (L p. 297). In the event, it would be the end of 1961 before *Pale Fire* was with its publisher and the index cards on which it had been written were deposited in the Library of Congress. Between his confident letter of 1957 and his pessimistic one of 1959, *Lolita* had been published in America, with tremendous success: this presumably freed its author from the financial necessity of producing another novel quickly (and doubtless enabled him to repay the advance effortlessly), but seems to have occasioned a period of uncertainty for Nabokov, who in 1959 was also making up his mind whether or not to attempt a screenplay for Stanley Kubrick's film of *Lolita*, as well as deciding to leave his job at Cornell and contemplating residence in Europe. Although the exiled king and his clumsy assassin remained along with the waxwings, the differences between the novel projected in 1957 and the novel published in 1962 are marked: the audacious form of poem and commentary, the character of John Shade and Kinbote's flagrant homosexuality, are all significant alterations – born perhaps in the air of freedom Nabokov breathed once he had made up his mind about these other issues. Inaccurately billed as 'his first new novel since *Lolita*', *Pale Fire* enjoyed a sojourn in the best-seller lists – somewhat to its author's surprise – which was doubtless due to its forerunner's notoriety rather than to the reading public's appetite for experimental fiction. Interestingly, in his 1957 letter Nabokov wrote that he had considered calling *Pale Fire The Happy Atheist*: a title which signalled his sense of its taboo-breaking.[2] Albeit that the novel he actually published was different, we can see that, taken together with the preceding *Lolita* and the following *Ada* (1969), *Pale Fire* describes a rising arc of illicit sexual behaviour, from Humbert's nympholepsy, through Kinbote's pederasty, to Van and Ada's incest.

The extraordinariness of the relationship between Humbert and Lolita is echoed, in *Pale Fire*, in the unlikely pairing of prosaic Shade and fanciful Kinbote, further complicated by the fact that the former is the poet and the latter his commentator. The essentially bipartite construction of *Lolita* is extended in the quadripartition of *Pale Fire* into Foreword, Poem, Commentary, and Index: a literary nonesuch

2. His afterword to *Lolita* suggested 'the total atheist who lives a happy and useful life' as one of three basic taboos for American publishers.

in which the novel masquerades as a scholarly edition, and gives the sharpest point to Nabokov's declaration that one of the functions of all his novels was to prove that 'the novel in general' does not exist. It remains one of the few novels to present readers with a serious problem not merely as to meaning, but as to the very sequence in which to read it: should we start at its physical beginning and proceed page by page to its end, or not? Indeed, leaving aside the existence of a 'first' page and a 'last' page, it is not easy to determine just where this story begins, nor where it ends. Surprised as we may be to find a 'novel' consisting of poetry, there are some precedents: Elizabeth Barrett Browning's *Aurora Leigh* is one, although the more relevant example is of course *Eugene Onegin*. The closest model for the unbalanced mixture of poem and commentary in *Pale Fire*, however, is Nabokov's own edition of Pushkin's verse-novel: in which the 250 pages devoted to his literal translation are accompanied by nearly 1200 pages of scholarly apparatus (in addition to this there is an index of over 100 pages).

Unlike *Lolita*, *Pnin* or *Ada*, it is not immediately obvious why *Pale Fire* should have this title. It is the name given by its author, a New England poet called John Shade, to an uncompleted poem of 499½ heroic couplets. The book purports to be a scholarly edition of this text; whose foreword, copious annotation and index are the work of Shade's erstwhile neighbour in New Wye, who represents himself as the poet's friend and confidant, and identifies himself as Dr Charles Kinbote. His foreword reveals that Shade has been killed in mysterious circumstances which have left Kinbote in possession of the poem's manuscript; and that he has resisted attempts made by publishers, fellow-academics and by Shade's initially compliant widow, to limit his total editorial control over the material. Our peaceable scholar has found it necessary to seek out a 'new incognito in quieter surroundings', but unfortunately for him the 'wretched motor lodge' in Cedarn, Utana where he prepares his edition, has turned out to be adjacent to an amusement park. Notwithstanding the presence of such distractions and the absence of any research facilities, Kinbote manages to plump out the 29 pages of his dead friend's poem with 175 pages of commentary; which represents his effort to impose upon 'Pale Fire' the subject he had imagined – as a result of pumping Shade during its composition – that the poem would undertake. As his notes progress, their relation to the textual actuality of Shade's poem grows ever more tenuous; they lose referentiality and take on their own wild life as the commentary becomes what Wallace Stevens said a poem is: 'the cry of its occasion, / Part

of the res itself and not about it'.[3] But keeping pace with this bur-
geoning fantastication, both intrinsic and inimical to it, is the imagined
approach of the unimaginative assassin Gradus: who comes into ever
closer focus the further Kinbote unrolls the elaborate magic carpet of
his notes.

Shade's poem, which we are led to believe would have consisted
of 1,000 lines symmetrically arranged in four cantos, is a versified
meditation on this life and any possible afterlife, on his relationship
with his wife Sybil, and on their continuing grief over the apparent
suicide of their ungainly daughter, Hazel. Disappointed Kinbote,
in his note to the non-existent line 1,000, dismisses the whole as
'an autobiographical, eminently Appalachian, rather old-fashioned
narrative in a neo-Popian prosodic style'; and it is certainly true that,
despite occasional felicities, Shade's verses never achieve the sustained
supple precision of his admired Alexander Pope. But something other
than artistic dissatisfaction underlies Kinbote's comment: his disap-
pointment stems from the absence in 'Pale Fire' (for which he had
suggested the title 'Solus Rex') of all traces of the stories he had
shared with – or inflicted on – Shade, concerning Kinbote's distant
homeland of Zembla, and the trials and adventures of its deposed
monarch, Charles the Beloved. *Pale Fire* is his increasingly audacious
attempt to interpret 'Pale Fire' in the light of his own *idées fixes,* and
to produce from it a different kind of autobiography. As literary theory
teaches us, the author is dead; and with Shade safely out of the way
Kinbote can, in his foreword, embark on his appropriation of the
poem:

> Let me state that without my notes Shade's text simply has no human
> reality at all since the human reality of such a poem as his (being too
> skittish and reticent for an autobigraphical work), with the omission
> of many pithy lines carelessly rejected by him, has to depend entirely
> on the reality of its author and his surroundings, attachments, and
> so forth, a reality that only my notes can provide. To this statement
> my dear poet would probably not have subscribed, but, for better or
> worse, it is the commentator who has the last word. (p. 25)

Just as Lucifer was too proud to serve in Heaven, so this startling
commentator refuses to subordinate himself to any of his dear poet's
intentional fallacies, but instead constructs an *apparatus criticus* during
whose course he metamorphoses from the dull grubbishness of editorial
servitude into the self-sufficient splendour of kingliness (a 'monarch'

3. This comes from his poem 'An Ordinary Evening in New Haven'.

is an American butterly). Deposed elsewhere, Kinbote sets up a government in exile in 'Pale Fire', and affirms himself the hero of Shade's story.

The relationships between the main characters echo those of *Lolita*: as sexual invert and imperial fantasist, Kinbote resembles Humbert, while Shade, as representative of the ordinary universe and Kinbote's emotional object, has something in common with Lolita; his wife Sybil interposes herself much as Charlotte Haze did, and Gradus the destroyer has the same negative force as Quilty. Thematically, there are some congruities with the intervening *Pnin* (whose eponym reappears here as a 'farcical pedant' in charge of the Russian Department at Wordsmith): Shade inherits Pnin's heart condition, while Kinbote has a prying eye; ideas of exile and displacement abound, and *Pale Fire* continues the earlier book's satirical observations of life at a provincial American university. Much of the humour comes from its depiction of the awkward foreigner Kinbote and his apparent unawareness of the impression he makes on his new colleagues – as in this recollection of his first meeting with Shade in the refectory:

> I was invited to join him and four or five other eminent professors at his usual table, under an enlarged photograph of Wordsmith College as it was, stunned and shabby, on a remarkably gloomy day in 1903. His laconic suggestion that I 'try the pork' amused me. I am a strict vegetarian, and I like to cook my own meals. Consuming something that had been handled by a fellow creature was, I explained to the rubicund convives, as repulsive to me as eating any creature, and that would include – lowering my voice – the pulpous pony-tailed girl student who served us and licked her pencil. Moreover, I had already finshed the fruit brought with me in my briefcase, so I would content myself, I said, with a bottle of good college ale. My free and simple demeanor set everybody at ease.

Humour notwithstanding, there is a sort of twisted exuberance in Kinbote's verbal rococo with its pedantic exfoliation of detail; together with a naive vulnerability – or kingly disdain – amidst these fat professors and their potential mockery (Shade's carnivorous challenge may have concealed the litmus test of an anti-Semite).

This inauspicious meeting has taken place on 16 February 1959, eleven days after Kinbote had arrived to take up a lectureship at the university where Shade, a minor poet of regional significance, has the order of his magnitude enlarged by being its distinguished writer in residence. Kinbote in his lilac slacks moves into Judge Goldsworth's house, installing his fast red car in its garage and twin ping-pong tables in its basement, as an erotic adjunct. He then pays court ('lays siege' might be more accurate) to his relatively famous neighbour: from which several strands of comedy unwind. The first lies in the unconscionable assiduity with which an intrusive academic nobody seeks to impose his company on a literary (stuffed) lion, hoping perhaps to make a career out of the crumbs dropped from the Master's table. He desires to install himself as the preferred disciple, and is necessarily jealous of other influences – especially of the poet's wife, who intercepts him, for example, as he officiously takes up to her husband the junk mail he himself spurns to read. The second strand lies in the disproportion between the acolyte's adulation and the evident mediocrity of the object idolized: Shade is treated with a fawning sycophancy that would be excessive were he Pope (whether pontiff or poet); his not very *bons mots* are preserved by Kinbote with the same ecstatic devotionalism that earlier ages lavished upon saintly relics; as when Shade declares, of 'a certain burly acquaintance of ours: "The man is as corny as a cook-out chef apron." Kinbote (laughing): "Wonderful!".'. Such a scene, with the faithful disciple applauding at his master's elbow, reminds us that sycophancy is not unconnected with egotism; and a further comic strand unspools as arrogance displaces adulation, and the Master is shouldered to the margin of the scene, then off-stage altogether. The disporportion between the two men as individuals provides another comic aspect: Shade is fat, scruffy, married and heterosexual, a happy atheist rooted in New Wye, where he has lived always in the same house; Kinbote is tall, clothes-conscious, divorced, deeply homosexual and unhappily religious, in remote exile from Zembla or the Lord knows where – 'a citizen of somewhere else', to borrow Hawthorne's self-description.

Installed next-door, Kinbote subjects his celebrated neighbour to the closest scrutiny, extravagantly confirming Jane Austen's perception that living in society is much the same as dwelling in a neighbourhood of voluntary spies. When in his foreword he talks of 'seeing more and more' of Shade, he refers not to social intercourse but to the 'orgy of spying' on which he embarks. From various vantage-points in house and garden, Kinbote observes the phases of Shade's days and nights, revealing not only his own extraordinary nosiness and manic solicitude for the poem under composition, but also his capacity for misinterpretation: thus a repeated scene he had (hopefully?) supposed to be the evidence of marital discord, with Sybil flouncing from the room, turns out to be her response to the telephone. Much more or less conventional humour derives from Kinbote's conviction that Shade desires his friendship, but is prevented from enjoying it by his termagant wife and other envious interposers. To meet such obstacles Kinbote resorts to elaborate stratagems, both in imagination ('What would I not have given for the poet's suffering another heart attack . . . leading to my being called over to their house, all windows ablaze, in the middle of the night, in a great warm burst of sympathy, . . . a resurrected Shade weeping in my arms ("There, there, John")') and in actuality: he discovers the remote and secret location where they plan to take their summer holiday, and books himself a nearby cabin ('The more I fumed at Sybil's evident intention to keep it concealed from me, the sweeter was the forevision of my sudden emergence in Tirolese garb from behind a boulder and of John's sheepish but pleased grin'). The Shades, of course, never take that holiday; but it is in this cabin that Kinbote will prepare his edition.

During the five months of their vicinity Kinbote is thrice invited to dine with the Shades, who on each occasion overlook his vegetarianism. For their parts, of the 'dozen or so' invitations with which their neighbour bombards them, the Shades accept an equal three. Instead of the colleagues he had met *chez eux*, Kinbote evens up numbers with first, a male student (probably a catamite), second, his negro gardener, and third, a 'stunning' female student in a leotard (this may be his revenge, along with the vegetarian fare he inflicts: this blonde is rumoured to excite Shade's amorous propensities, and husband and wife leave ten minutes after her arrival). Kinbote's friendly feelings toward Shade are obviously unreciprocated; but we also see his pitiable loneliness and his social insignificance, in the meagre muster of this guest-list. Betrayed – as he sees it – by his male lovers, ridiculed by his colleagues, frightened of the dark, and subject to delusions that somebody is out to get him (perhaps the consequence of unkind practical jokes played on him by a disaffected sexual partner, see his note to line 62), Kinbote is well aware of the difficult margins he treads, and of the humiliations implicit in his futile dependence on the kindness of strangers. As a gauche oddball attaching himself to the campus's local luminary, he attracts much mirth – his foreword remembers the faculty wife who snickered at him, as, after her 'dreary' party, he was 'helping the tired old poet to find his galoshes' (there really are people like this). In his notes Kinbote confronts the reported contempt of Sybil, who was wont to describe him as the 'monstrous parasite of a genius'; and he rises to the occasion by the comic grandeur of the forgiveness he offers 'her and everybody' (although his edition has a sting, literally in its tail: the index disdainfully omits his petty persecutors, and dismisses Sybil with a resonating curtness: 'S's wife, *passim*').

Parasitism, the condition of people who take without giving, would seem to be a central concern of a book whose title comes from *Timon of Athens* (as Kinbote overlooks, having only his Englishing of a Zemblan translation to go by):

> The sun's a thief, and with his great attraction
> Robs the vast sea; the moon's an arrant thief,
> And her pale fire she snatches from the sun;
> The sea's a thief, whose liquid surge resolves
> The moon into salt tears; . . . each thing's a thief.
>
> (IV. iii. 434)

The misanthropic Timon here discerns creation as a system of cosmic larceny. In the context of the novel, it is natural to suppose that the commentary is to the poem as moon is to sun in this analysis, a mere reflector of light elsewhere generated. But it is interesting to note that even in Shakespeare's image the moon is not a passive recipient, but actively 'snatches' her illumination; and as we read into Kinbote's commentary, any sense of its origination in Shade's poem is speedily eclipsed by independent glintings: most obviously its Zemblan elements, but also such items as the hilarious playlet Kinbote improvises around the ghost-watch in the barn (see his note to line 347). Owing its title to Shakespeare, *Pale Fire* is prefaced by an epigraph from Boswell's *Life of Samuel Johnson*; Shade is said physically to resemble Dr Johnson, and at times his mode of speech as reported by Kinbote is Johnsonian (see especially the note to line 894). This epigraph, concerning Johnson's solicitude for his cat Hodge, perhaps echoes Nabokov's amused tenderness toward his creature Kinbote who, like Hodge, is not to be shot; but it is also relevant to recall that Boswell's biography is a celebrated example of writing which transcends its function of being 'about' something external to itself, to emerge as a work of art in its own right.

Notwithstanding Shade's physical resemblance to the great lexicographer, it is perhaps Kinbote, with his dark sense of human sinfulness only partially relieved by his Christianity, and his nocturnal terrors ('solitude is the playfield of Satan'), who is spiritually closer. Shade's mental climate seems more suggestive, in a minor way, of the positivistic Pope of the *Essay on Man* – a poem to which Shade's work makes more than one allusion, and which was memorably dismissed by Johnson ('Never were penury of knowledge and vulgarity of sentiment so happily disguised'). A similar stricture could perhaps be applied to 'Pale Fire', a poem which, albeit touched by one deep grief, ends (or almost does) with the affirmation of a tepidly resigned optimism. These are the annals of a mind which hasn't travelled very far; and even the would-be tender evocations of the poet's love for his wife are somewhat sullied by his statistical enumeration of their acts of sexual intercourse. The poem is written by a man sufficiently self-engrossed to imagine that his readers wish to know how he sets about shaving. On the other hand, Shade's attachment to the everyday is what has given him strength:

> And I'll turn down eternity unless
> The melancholy and the tenderness
> Of mortal life; the passion and the pain; . . .
> Are found in Heaven by the newlydead
> Stored in its strongholds through the years. (pp. 44–5)

The refusal to part with the local details of a life is what finally animates this anti-transcendental poem; his researches into a hereafter end in farce, and even his sense of a grand (if impenetrable) design is squashed, with the aid of his untranscendental wife, at the end of Canto Three. And so 'Pale Fire' affirms life as texture and process rather than meaning and purpose, a pattern of recurrence relaxedly embraced at its close:

> I'm reasonably sure that we survive
> And that my darling somewhere is alive,
> As I am reasonably sure that I
> Shall wake tomorrow, on July
> The twenty-second, nineteen fifty-nine,
> And that the day will probably be fine. (p. 58)

Shade's philosophy of reasonable assurance and probability is of course rebuked by the fact that his confidence in the proximate future turns out to have been misplaced: whatever that day's weather, he does not live to see it. 'Pale Fire' is a poem about failed transitions, failed metamorphoses: the bird that cannot fly into an imagined dimension, the daughter who does not outgrow her ugly duckling phase, the poet who will not live to finish his poem. The 'waxwing' that kills itself against the sky-reflecting window, identifiable from the description as a cedar waxwing (*Bombycilla cedrorum*), perhaps anticipates Cedarn, where Kinbote will compile his notes; but perhaps there is an additional allusion to other wings of wax: Icarus's, which melted when he flew too near the sun. Icarus's story is of the limitations of human ingenuity and the limits of human aspiration, and is relevant to *Pale Fire* which, like *Lolita*, is aware of imprisonment: 'We are most artistically caged', declares Shade's poem, and like Sterne's caged starling with its cry of 'I can't get out, I can't get out', our utmost artistry consists in describing the state of our imprisonment, or in drawing the bars of our cage.[4] This is a continuing theme in Nabokov's fiction; as we read in *Ada*:

> Are we *really* free? Certain caged birds, say Chinese amateurs shaking with fatman mirth, knock themselves out against the bars (and lie unconscious for a few minutes) every blessed morning, right upon awakening, in an automatic, dream-continuing, dreamlined dash – although they are, those iridescent prisoners, quite perky and docile and talkative the rest of the time. (p. 100)

If Shade's 'Pale Fire' is a poem whch declines to attempt Icarian flight, its commentator has fewer qualms, and quickly leaves the *terra firma* of the text behind: perhaps because moonlight offers no threat to his flying (which, after all, is done by means of what Keats termed 'the viewless wings of Poesy'). His colonization of Shade's poetic territory (embarked upon in the foreword when he suggests we read his notes first, then poem and notes together, and then the notes again!) gathers momentum as his friend's Appalachian tangs are overlaid by Kinbote's costume-drama of Zemblan derring-do. For underlying the comedy of literary-university manners between Shade and his editor is an altogether more stirring tale: we are to believe that the quasi-Ruritanian status quo in the remote northern kingdom of Zembla has recently (May 1958) been overturned by a Soviet-inspired political coup; the King, Charles 'the Beloved', has escaped from captivity and, aided by dashing and loyal noblemen, has after suitably picturesque adventures succeeded in fleeing the country; the new regime, however, is anxious lest he become the focus for émigré dissidence, and an assassin is despatched to track him down, whose progress is synchronized with various stages in the composition of 'Pale Fire'. As the commentary unfurls and Kinbote expatiates upon the various manly customs of his native state and incidents in the King's life, his boldness increases to the point where we realize that he is no Zemblan patriot-in-exile but the King himself: the note to line 691 finally topples over into the first person rather than the third. He uses his love of literature to pose as a humble lecturer, and his love of Zembla to impose on Shade the task of writing the commemorative epic his homeland deserves. But Gradus the assassin,

4. This comes from *A Sentimental Journey* (1768); the starling is alluded to by Humbert in his poem about Lolita, and explicated by Nabokov in his lectures on *Mansfield Park*, where Jane Austen also alludes to it.

whilst stupid, is implacable; and by a series of clumsy accidents is able to trace King Charles to New Wye where, in a final stupendous botch, he misses his target and kills instead his companion, John Shade. The King, sure that the poem will turn out to be the one he would himself have written, hides the manuscript; and then perceives the necessity of hiding himself (Gradus, of course, denies all knowledge of Zemblan politics).

This, straightened out, is what Kinbote's meandering and disjointed narrative amounts to; this is his version of 'Pale Fire': a blatant extravaganza simultaneously redolent of *The Prisoner of Zenda*, a Cold War spy thriller and a High Camp fantasy, enlivened by a vigorously opinionated humour and set off by intermittent glimpses of a darkened and suffering spirit. The contrast between this and John Shade's tranquil commonsensicality could hardly be sharper: Nabokov has constructed the widest possible gulf between text and interpretation, between Shade's writing and Kinbote's reading. There is, as Mary McCarthy was the first to point out in her review of *Pale Fire*, a sequence of events to be deduced, other than the one Kinbote asserts. Shade has indeed been shot, not by a secret agent but by a homicidal maniac (see note to lines 47–8) in mistake for Judge Goldsworth, who had sentenced him; that Shade and Goldsworth look alike has been established by their mutual resemblance to an ugly old woman who works in a university cafeteria. Because Shade was accompanying Kinbote back toward the Goldsworth house the killer, who had escaped from an asylum for the criminally insane, assumed he was the Judge. Kinbote was marginal to the entire affair; but he has the manuscript which, although initially it disappoints him, offers the chance of establishing his fantasy as fact. Although his first reading of 'Pale Fire' makes him feel, one supposes, like an Ancient Mariner who has singularly failed to hypnotize his hearer with his tale, increasingly coercive re-readings show him the poem's Zemblan intratextuality, which his editorial additions enable him to explicate (it is possible that the marginal glosses which Coleridge added to his *Rime of the Ancient Mariner* offered Nabokov a distant model).

Just as Kinbote educes his version in defiance of the apparently uncooperative text of 'Pale Fire', so we construct an interpretation of *Pale Fire* that is at variance with the story that he tells us. In our version, Kinbote suffers from delusions (Nabokov described him, in letters, as a madman); but since the book is composed of his structure of reality as well as Shade's, theirs have to be seen as complementary, rather than contradictory, accounts. Who, then, *is* Kinbote? Some interpreters have seen him as Shade's projected *alter ego*, others have seen Shade as Kinbote's *alter ego*; the most plausible answer is that he is in 'reality' Professor Botkin, who is mentioned in the text (most significantly in the note to line 172), and is described in the index as 'an American scholar of Russian descent' (interestingly, this index entry omits mention of line 172). Botkin has confected the persona of Kinbote/King Charles for reasons which are hinted at in a conversation he intrudes upon between Shade and Mrs Hurley concerning, they tell him, a local railway employee who believed he was God and started redirecting the trains (how does one redirect trains without shifting the tracks?); Shade hails him as a fellow poet, arguing against her simpler label 'loony' that 'one should not apply it to a person who deliberately peels off a drab and unhappy past and replaces it with a brilliant invention. That's merely turning a new leaf with the left hand'. From our perspective, this quite obviously applies to the left-handed Kinbote (to retain this name), who has elaborated the realm of Zembla – much as Humbert elaborates his never-never land of nympholeptic bliss – where he can be beloved King rather than unloved outsider (one of Zembla's towns is called 'Yeslove').

'Zembla' is alluded to in Pope's *Essay on Man*, where it denotes 'Nova Zembla', two long islands in the Arctic Circle which now form part of the territories of the Soviet Union, and are given on English maps as Novaya Zemlya. Pope uses it to illustrate the relativity of human definitions:

> Ask where's the North? At York, 'tis on the Tweed;
> In Scotland, at the Orcades; and there,
> At Greenland, Zembla, or the Lord knows where.
>
> (Ep. ii, ll. 222–4)

Kinbote, in his gloomy annotation of line 937 of 'Pale Fire', records that Shade had made a marginal jotting of the third of these lines in his manuscript. It is interesting to note that even if Pope's 'Zembla' represented an actual place, it nonetheless occurs as a gradation toward the fantastic; for Kinbote's 'blue inenubilable Zembla' is a logical step further into the irreality of a land of lost content, a Lord-knows-wheredom in which such a one as he could have found happiness (even the adjective he uses to describe it is his own invention). It has its similarity to the land beyond the looking-glass – Zemblan, we are told, is 'the tongue of the mirror' – which accounts for the theme of duplication, but also recalls the 'nonnons' of which Cincinnatus's mother told him, in *Invitation*: these are grotesque-looking objects, whose misshapenness is rectified when reflected in a suitably distorting mirror. Zembla becomes for Kinbote a glass in which those aspects of his character which render him marginal or ridiculous in actuality are effortlessly reflected as the lineaments of majesty. His friendlessness and his pederasty find themselves transfigured: instead of the evident difficulties he experiences in New Wye in finding trustworthy ping-pong partners, in Zembla, manliest of states (with its pink-coned and phallic Mount Falk, and its Béra Range, 'an erection of veined stone and shaggy firs'), homosexuality is the national pursuit; nude willing boys throng the regal antechamber, and the King is surrounded by valiant retainers devoted to his well-being.

So Kinbote, out of his estrangement from what he dismisses as 'our cynical age of frenzied heterosexualism', weaves his homoerotic idyll: in which there are some charming elements, such as the young King's journey through the secret passage with Oleg, Duke of Rahl, in which a *Boy's Own* adventure story is crossed with a parody of homosexual soft-core kitsch; or the note to line 408, where Gradus's snooping visit to a Swiss villa is punctuated by Kinbote's homoerotomanic fantasy of a bronzed youth whose lithe nakedness is emphasized by a succession of scant coverings, the last of which (a 'Tarzan brief') is doffed. But just as Humbert, even in the security of his own narrative control, could not utterly expunge the variant voices and divergent visions of those also involved in the story, so Kinbote, even on the Zemblan throne, is prey to revolution, and moreover cannot shake off a sense of guilt and self-disgust. Although he can effortlessly resist the importunate nudity of Lady Fleur when she tries to ensnare him, his feelings toward Disa, Duchess of Payn, whom he is persuaded to marry in vain hopes of fathering an heir, are complex and tormented (see her entry in the index). His apparently unrepentant evocations of Zemblan ingledom are accompanied by his awareness of the tawdriness of his pederastic inclinations, of his loneliness and unhappiness. If we look beyond the Zemblan rigmarole to the unfortunate soul who has begotten it, we may suppose that the whole fantasy amounts to an attempt to compensate Botkin for the land and language he has lost, the wife he has disappointed, and the betrayals he has suffered: the brilliant invention that effaces his drab and unhappy past. At what stage his private day-dream accrued the power and necessity of an obsessive delusion which he attempts to impose on others, we cannot know: but we may well suspect that one act of perfidy too many tipped the balance, and that Gerald Emerald, who is throughout perceived as a sort of minor Judas, was responsible (see index entries for lines 741 and 894 under 'Kinbote').

In 'Pale Fire' Shade quotes a line from *Essay on Man*, and in the note (to ll. 417–21) Kinbote also gives Shade's variant. It is worth quoting a little more of Pope's verses, which is where Shade found the title for his book on that poet:

> See the blind beggar dance, the cripple sing.
> The sot a hero, lunatic a king;
> The starving chemist in his golden views
> Supremely blest, the poet in his muse.
> See some strange comfort every state attend.
>
> (Ep. iii, 267–71)

Pope here expounds his idea that however miserable our condition, we never wish to be anybody other than ourselves; he deduces from this the notion that inherent in each life is the compensation for its own misfortunes. Shade quotes the first of these lines in 'Pale Fire', and finds it 'vulgar' (although Pope, who was severely disabled, surely acknowledged affinity with the cripple as well as the poet); but it is clear that the line following, which he decided not to quote, seems to describe Kinbote: the 'strange comfort' of Kinbote's lunacy is that it enables him to imagine himself as king (and bibulous Shade perhaps exemplifies the sot as hero). What we could call the theory of compensation in *Pale Fire* goes further: Shade is physically awkward, but intellectually graceful in the run of his verses; Kinbote declares of him that 'his misshapen body, that gray mop of abundant hair, the yellow nails of his pudgy fingers, the bags under his lusterless eyes, were only intelligible if regarded as the waste products eliminated from his intrinsic self by the same forces of perfection which purified and chiseled his verse'.[5] This contrast that is as well a complementarity, between Shade's own ungainliness and the poise of his poetry, is also to be found in the relationship between the everyday-ness of 'Pale Fire' and the extraordinariness of Kinbote's moonlit commentary on it.

But, Pope's argument notwithstanding, compensation is not a universal principle: Hazel Shade, for example, seems to refute it, having found nothing worth staying alive for. In Kinbote's case, *Pale Fire* is open-minded about whether the pleasures of his predicament outweigh its pains: he has an idiosyncratic and active imagination, to be sure, but as Dr Johnson noted, all power of fancy over reason is a degree of insanity. The fixity of his vision leads our editor to mistake certain details – for example, he misses the point about the

---

5. I am struck by the similarity between this and the tribute of a recent biographer-obituarist recalling his first meeting with his future subject: 'Even then he was not slim, yet he carried about him an atmosphere of thinness' (*The Independent*, Thursday 4 April 1991, p. 3).

sports headline saved by Shade's Aunt Maud (see line 98 and his note), in which 'Chapman's Homer' refers to the winning home run completed by a player called Chapman: an instance in which Kinbote's ignorance of baseball mirrors the sports-writer's ignorance of Keats's sonnet. More significant, however, is the insensitivity which prevents Kinbote from fully measuring the depth of the Shades' grief over their daughter's suicide. His self-absorption leads him to interrupt when Shade seemed about to speak of Hazel at the site of the old barn, where the poet's memory had been prompted by the specifics of place (note to line 347). Significantly, Kinbote banishes any tender recollections on his part by expounding some aspect of Zemblaniana: willing as ever to substitute fantasy for fact, faraway places for actual earth, and his own story for Shade's – just as he uses the poem's evocations of Sybil as the occasion to set forth the circumstances of the marriage between King Charles and Disa. His homosexuality was presumably allotted him by Nabokov not only as an appropriate social handicap (these are the 1950s), but also as a logical consequence of the narcissism that would make the world a mirror for the self. Yet these facets of his character are linked with the creative excesses of his commentary that exuberantly displaces the poem it was meant to illuminate, and with his narrative urge (which is also a kind of will-to-power) that does not cease, even in the index. Whether trusting that his reader has enjoyed a note (the one with farmer Griff and his daughter Garh in it), or uttering his splendid invective against swans in the note to line 319, Kinbote, the man who can imagine a funfair where there is only a radio, exemplifies the festive extravagance as well as the self-enclosure, of the mind that is its own place.

'I trust the reader appreciates the strangeness of this' declares Kinbote, appropriating Shade's tribute to Sybil for his own Queen Disa, 'because if he does not, there is no sense in writing poems, or notes to poems, or anything at all'. The expanding freestyle of Kinbote's 'strangeness' is antithetical to Shade's centred homeliness and the predictable returns of his rhyming couplets; but Gradus, who takes ever-clearer shape as the commentary progresses, and whose approach is synchronized with the composition of 'Pale Fire', is the negation of each. Both poet and commentator, in their different ways, exemplify the life of the mind (Kinbote asserts that 'Mind is a main factor in the making of the universe', in the note to line 549); but Gradus is an emanation from the world of impure matter, most memorable for his diarrhoea. He is the emissary of an organization whose members call themselves the Shadows, and is reminiscent of the agents of death sent out from Soviet Russia, particularly under Stalin. As moon reflects sunlight, so poem mirrors world, and commentary poem; but whereas a reflection is the retransmission of light, a shadow is its absence: light-cancelling Gradus is a man without art, for whom 'Browning' denotes a gun and not, as for Kinbote, a poet. His moral and political philosophy is reminiscent of Ekwilism, glorifying the general and abhorring the specific. Yet for all his contemptible apishness and his clumsiness, Gradus is inevitable: like a reflection, a shadow requires light for its existence, and he represents the destructive underside of creation, the negative that is called into being by the very existence of a positive (just as, by building a house, we endow it with the capacity to fall down). Exactly the same age as Kinbote, he could be taken to be his obverse twin; but whether as the programmed assassin of Kinbote's fantasy, or as the vengeful lunatic of the police account, he is lethal, and Shade dies.

Yet Shade's death is as necessary to the emergence of *Pale Fire* as Lolita's was to *Lolita*: it leaves Kinbote in possession of the manuscript, by means of which he becomes the shadow of the slain waxwing, and embarks on his prolonged flight of fancy. In being an accident, conforming neither to the intentions of the lunatic nor to those of Gradus, Shade's death is a piece of pattern-breaking, of topsy-turvical coincidence, that sets in motion matters in a different dimension from the predictable. 'Pale Fire' alludes to Goethe's poem *Erlkönig*, in which a father, riding by night through stormy woodland, tries to calm his child's mounting fears that he is about to be abducted by the devilish erl-king (it has been memorably set to music by Schubert: a fine example of pale fieriness). This three-cornered dialogue, involving the exchanges between father and son, and son and erl-king, also implies a dialogue between poet and reader: just as *Pale Fire* is a three-hander between Shade, Kinbote and Gradus (Shade's name in Spanish is the name of a card-game for three players, 'ombre'), as well as a game between Nabokov and his reader. In *Erlkönig*, the parent's commonsensical answers to the child's increasingly frantic questions have no effect ('you can't see the devil, that's just mist; you can't hear him, that's wind in dry leaves'), and when they reach the farmhouse his son is dead in his arms. Even if the father did not believe in the erl-king, the child did: whatever your imagination lends itself to becomes real for you – and within the poem the erl-king's voice has the same status as the father's. *Pale Fire* obviously contains different levels of imagination, like Goethe's poem; but interestingly, it is Shade, the representative of common-sense, who dies, whereas Kinbote, devotee of strangeness, survives.

He may not survive for long, however. Nabokov declared in an interview that Kinbote commits suicide as soon as his editorial work is finished; and, in spite of his closing assertion about continuing to exist, in his note to line 493 – a discussion of suicide 'with himself', as the index interestingly puts it – it may not be accidental that the room numbers in the skyscraping hotel from whose window Kinbote imagines rolling are 1915 and 1959: which would be the dates on his tombstone, were he to act on his creator's hunch. As his commentary achieves its apotheosis and annotates an unwritten line, and as his fantastic effort declines, having wrenched from 'Pale Fire' the book in himself whose pages Shade was to cut (as he puts it), Kinbote confesses that 'my notes and self are petering out'. For if Shade's text has no human reality without his notes, *he* has no human reality without Shade's poem. In fact, *Pale Fire* consists in the wild interplay or counterpoint between their visions and voices: its dynamic depends on the tension between them, and their relationship, for all its comic distortions, is one of symbiosis rather than parasitism. Kinbote, who was at best a human curio to Shade, institutes himself as the poet's bosom-companion and soul-mate during the course of his notes, and finally produces Shade as a kind of secretary or amanuensis, writing *his* book. The 'secret' of Kinbote's that Shade thinks he has guessed is not, of course, that he is the King, but that he is a queen; and yet his Zemblan dreamery is perhaps more interesting, more creative, and more important than his sexual orientation, just as his variations may finally be more imaginative than Shade's theme.

To 'explain' Kinbote's character in terms of his homosexuality would be to engage in a pseudo-Freudian analysis which *Pale Fire* has parodically pre-empted. Like *Lolita*, it has resemblances to the mystery novel or whodunnit, yet any answer we can give as to what 'really' happens is much less rich than the whole of what goes on, and flies in the face of the overt and absolute fictionality of the entire work. The commonsensical father insisted there was no such thing as the erl-king, but his son died nevertheless; and we as readers lend our imaginations to the poem in which this happens. To look for theme at the expense of variations is to decline the invitation *Pale Fire* extends: when Horatio exclaims of Hamlet's behaviour that it is 'wondrous strange', the Prince retorts 'And therefore as a stranger give it welcome'; Shade's orderly universe co-exists with Kinbote's extraordinary one. It is no accident that the book's epigraph from Boswell is a digression ('This reminds me of the ludicrous account . . .'), itself catching the stern Dr Johnson in a moment of digressive fondness for Hodge; for *Pale Fire* celebrates the nonconformity of a discourse where things marginal displace the central subject, and Kinbote edges Shade aside (as in Shade's index entry). In his comment on line 810, where Shade predicates the finding of a 'web of sense' behind existence, Kinbote quotes the following, from a book lent him by his Cedarn landlord; in which the writer speculates on whom he would most like to meet in the afterworld, and chooses Aristotle:

> What satisfaction to see him take, like reins from between his fingers, the long ribbon of man's life and trace it through the mystifying maze of all the wonderful adventure . . . [*sic*] The crooked made straight. The Daedalian plan simplified by a look from above – smeared out as it were by the splotch of some master thumb that made the whole involuted, boggling thing one beautiful straight line. (pp. 205–6)

He offers this as similar to Shade's line of thought at that point of the poem; but if 'Pale Fire' attempts to straighten things crooked, *Pale Fire* is by contrast a maze whose boggling involutions are in themselves the point of the experience. If we were to offer a quotation as appropriate to the novel as Kinbote believes his to be to the poem, we could hardly better this, from Sterne's *Tristram Shandy* (words which Nabokov cited in a note to *Eugene Onegin*): 'Digressions, incontestably, are the sunshine;– they are the life, the soul of reading;– take them out of this book for instance,– you might as well take the book along with them'.

Demonstrating digression as sunshine rather than as moonshine is the function of *Pale Fire*'s triumphant display. Kinbote's editorial extravaganza is an example of a large-scale digressive tendency, but at the level of language smaller-scale deviations show what unexpected harvests can accrue from seeds which fall outside their designated furrow. 'Life Everlasting – based on a misprint!' is Shade's wry comment on the mountain/fountain crux which had failed to corroborate his heart-stopped vision of a mystical beyond; but read another way – and *Pale Fire* heightens our awareness of other ways of reading – his words suggest how misprints usher in, if not life everlasting, then at least a spontaneously-generated unexpectedness that has its own subversive quiddity. The best example of this is the remarkable series of near-puns which Kinbote gives in his note to line 803, citing a newspaper report of the Tsar's coronation which stated that a 'crow' had been placed on his head, apologetically amended the next day to 'cow'. Kinbote marvels that the English sequence crown/crow/cow should reproduce the degrees of difference of their Russian equivalents, but leaving this extra dimension aside, the consequence of these small errors is beguilingly to summon up a solemn ceremony, in which with escalating oddity first a black bird and then a large mammal are placed, with dignity, upon the new monarch – uneasy lies the head that wears a cow!

It is the disruptive creativity of this departure from intended sense that enacts, in miniature, the more substantial unconventionality of

Pale Fire: a king with a crown on his head is an inert image; a king with a cow on his head is bizarrely interesting. This illustrates the transformative powers of language itself, its capacity haphazardly to generate meanings other than those intended; the implication is that these semantic accidents occurring through indirection are often more productive than the planned message, as unauthorized entities sputter into being (this is part of the point of 'word golf', in which words alter, letter by letter, into their opposites). Another example of this is to be found in Nabokov's early novel *Glory* (1932, tr. 1971), in which a misprision from the hero's childhood is recalled:

> This happened in the very same year when the grand duke of Austria was assassinated in a seraglio. Martin imagined that seraglio and its divan very distinctly, with the grand duke, in a plumed hat, defending himself with his sword against half-a-dozen black-cloaked conspirators, and was disappointed when his error became evident. (p. 18)

Substitute 'seraglio' for 'Sarajevo', and this is what might be engendered.[6] History, however, unrelenting, does not permit the substitution; and so instead of Martin's stirring scenario we have the episode which sparked off the Great War and all its carnage (a tragedy of errors which is memorably set forth in Rebecca West's monumental *Black Lamb and Grey Falcon*). Yet who would not prefer this picturesque digression to the terrible eventuality?

Just as Shade's sober poem begets Kinbote's pyrotechnics, so in spite of the latter's most ardent assertions we subject his account to a strong misreading, and deduce a different narrative from the tale he tells us. *Pale Fire*, with its index that presents the text as simultaneity, and with its playful disruption of the sequentiality of reading, makes us aware of the extent to which all reading is constructive, all books a function of the way we read them. Kinbote, with his monomanic determination to uncover in 'Pale Fire' the poem he wishes it to be, and with his insistence that the poet is his friend and that, really, he alone understands and appreciates the inner meaning, merely resembles most of us in an extreme dimension, and enacts on a grand scale the efforts of appropriation we ourselves embark on, as we seek to be wiser than their authors about the books we read and they wrote. As biography involves taking over someone else's life, so a certain type of literary criticism is a supreme instance of writing that attempts to displace the writing it's supposed to be about (in which the gentle reader is

6. It may be worth recording that as I transcribed this quotation I mis-typed 'the grand dyke of Austria', as if to extend the comedy of unintention.

disabused of any simple notions of a text, by a skilled interpreter who discloses its dark subconscious, the hidden motives gliding like sharks beneath the calm surface of apparent intention).

In his earlier novels *The Gift* (1938) and *The Real Life of Sebastian Knight* (1941), Nabokov had experimented with parodic forms of biography, and in *Pale Fire* he parodies the kind of biographer who secretly believes that in a just universe his subject should really be writing the story of *his* life, rather than the other way round – as well, of course, as parodying the kind of literary critic who conceives that his ideas about a book are much more interesting than what the author actually wrote. The genre most obviously parodied by *Pale Fire* is that of the scholarly edition, devotedly assembled by a self-effacing hermit whose only care is to establish his author's text: far from submitting to such harmless drudgery, Kinbote attempts to substitute his own work for Shade's disappointing poem. It is no accident that the book was written soon after Nabokov had finished his translation of Pushkin's verse-novel, for Kinbote's carefree arrogance with Shade's text is the obverse of his creator's pedantic literalness as a translator; his unfettered anecdotalism is the opposite of Nabokov's massive, unflagging and occasionally idiosyncratic scholarship, displayed throughout the volumes of his commentary on Pushkin. *Pale Fire* is the ludic counterpart to *Eugene Onegin*, whose scholarly punctilio it transmogrifies into editorial high-handedness. As we have seen, however, the parody is of other genres as well; for just as Kinbote changes from a madman to a king, so *Pale Fire* changes from poetry to prose, from whodunnit to cod biography to mock edition, in a comedy of transformations that includes the novel form itself.

Are such transformations evidence of Nabokov's quasi-religious metaphysics, prefigurations of an ultimate transcendent alteration – as when what was once a caterpillar emerges from its chrysalis as a butterfly? Or do they merely show the ways in which art is gloriously free where life is not? Asked if he believed in God, Nabokov gnomically replied, 'I know more than I can express in words, and the little I can express would not have been expressed, had I not known more' (SO p. 45). In his poem 'The Elixir' (subsequently, palely fierily, appropriated as a hymn), the seventeenth-century poet George Herbert wrote:

> A man that looks on glass
> On it may stay his eye;
> Or if he pleaseth, through it pass
> And then the heav'n espy.

Can the waxwing get beyond the window? Many of Nabokov's characters exist in what might be called a conditional mode, awaiting a completer being in some dimension they can imagine if not yet attain. Whereas earlier commentators were inclined to interpret the realm of aesthetics as the sphere, for Nabokov, in which the exasperated spirit could experience a version of completeness in the joys of free play (as opposed to the ball-and-chain of biological necessity), more recent criticism has begun to speculate that his poetry and fiction half-articulates a system of belief in an afterlife. His protests against time, and the endorsement of imagination and 'infinity of thought and sensation', uttered in *Speak, Memory*, assumed a different aspect when his widow Véra, in her foreword to a posthumous collection of his verse, claimed that 'the hereafter' (*potustoronnost'*) was the 'chief theme' of her husband's writing.[7] She asserted that every critic had missed this aspect, but subsequently W. W. Rowe made good the omission by demonstrating that a sense of a spiritual otherworld pervades Nabokov's work.[8] Just what we can do with such revelations is debatable; philosophically, it represents a knight's move of the sort that Nabokov delighted in, but at the level of fiction it remains the kind of proposition to which, in T. S. Eliot's terms, we can give 'poetic assent' even if not 'philosophical belief'. We can at least agree that believing in ghosts may not be so very different from believing in stories; and Henry James's ludic fable *The Turn of the Screw* comes to mind: for if we replace the Governess's ghost-story with the psychological rationale for her inventing it, we may overlook the deeply irrational urge that underlies our own appetite for fiction.

This returns us to issues of compulsion and playfulness, previously noted in the context of *Lolita*. At an early stage of composing *Pale Fire*, Nabokov had experienced tension between a contractual obligation to write it and its free development; and this tension between the legal and the imaginative is relevant to the finished novel. There is enormous fun to be had on its lexical playfields; but if 'lex' suggests 'word' in Greek, it denotes 'law' in Latin; what and where are the limits of wordplay? 'How far can you go?' is a question particularly appropriate to a book whose dominant image is that of a bird breaking its neck against an unseen barrier. When fictions begin to operate as realities, Dr Johnson noted, that is the beginning of insanity; and for all Kinbote's verbal inventiveness, his migraines and his sleeplessness suggest this darker side, his acknowledgement that dementia is also to be found in Arcady.

---

7. Quoted by David Rampton in his *Vladimir Nabokov – a Critical Study of the Novels* (Cambridge: the University Press, 1984,), pp. 98–9.
8. W. W. Rowe, *Nabokov's Spectral Dimension* (Ann Arber: Ardis, 1981).

For just as Humbert, writing in custody, found that he had only words to play with, so Kinbote towards the end sees how the limits of his language constitute the limits of his world, beyond which he cannot progress. Nabokov shows a sense of language being both power and prison – doubtless a consequence of being trilingual as well as being so gifted a writer but at the same time, so highly conscious a *user* of language. Such a sense can perhaps best be illustrated if we consider the word 'paradise': a word shimmering with beyondness, which denotes the ultimately blissful realm we hope to enter the other side of death; 'the abode of the blessed'. If we look it up in a good dictionary, however, we find that it comes from a Greek and earlier Old Persian word meaning a 'park'; and this discovery somehow lays bare the metaphorical, rather than the metaphysical, implications of that word: instead of the immortal otherworld, we have – however splendidly – an ornamental garden, an obviously human construct, a picture of an ideal place which refers us back to the world it was meant to have transcended. It isn't 'there', but here: rather than denoting another order of being after death, 'paradise' seems instead a human image projected onto the blankness of eternity, a human sound set against Pascal's silent, infinite spaces. It is the mirror we hold up between ourselves and our eventual extinction.

Yet this sombre reckoning is hardly the dominant tone of the *allegro sostenuto* of Kinbote's editorial activity. Language may finally imprison him, but it has nevertheless enabled him to extrude the comic wonderland of Zembla: its geography, its history and local colourings, his own regal ancestry, all are joyously, longingly elaborated; in seeming corroboration of the character in Pinter's play *Old Times* who declares that 'there are things I remember which may never have happened but as I recall them so they take place'. For *Pale Fire* may be a comedic inversion, not only of *Eugene Onegin*, but also of Nabokov's artful reconstruction of his own past in *Speak, Memory*. But if Kinbote's Zembla and Nabokov's pre-Soviet Russia represent for each a paradise lost, Nabokov has a much stronger sense than does his creature of the beneficial alternative offered by America, which exists beyond language. In common with the two preceding novels, *Pale Fire* has a responsiveness to and a responsibility toward its American locale, even if more markedly than in them this is offset by Kinbote's vision of a Zemblan Arcady; he nevertheless has glimpses, along with Shade, that in its way New Wye is a substitute Arcadia – and is in any case the only reality on offer. It is part of his pain that, like Humbert (and unlike Pnin), he cannot reconcile himself to living in the world that is available; yet his dissatisfaction with the given is the itch of his artistry – and as with Kant's dove which supposed it would fly more easily in a vacuum, Kinbote may disdain the local habitation in which he finds himself, but he needs its resistance: Zembla depends upon New Wye. On his first reading of Shade's poem, Kinbote misses that 'special rich streak of magical madness' he had hoped to have imparted, which he then confects by means of his commentary. This is doubtless another example of 'the madness of art', a distorting creativity that is celebrated in *Pale Fire*; but if ultimately it is a mirror rather than a window, it is nevertheless (in Wallace Stevens's words) 'A glass aswarm with things going as far as they can'.[9]

9. This is the last line of his poem 'Looking Across the Fields and Watching the Birds Fly'.

6

Conclusion

After *Pale Fire*, Nabokov published three more novels: *Ada* (1969), *Transparent Things* (1972) and *Look at the Harlequins!* (1974). These were the years of his late fame, when he was busy supervising translations of his works, and when he was sought out by interviewers willing to submit to the highly controlled 'dialogue' he instituted, whereby their questions were submitted in advance, and his answers were recited from scripted index cards. Nabokov acknowledged himself to be a poor conversationalist (he spoke English with a distinctly foreign accent), and this procedure suggests his lack of confidence with speech; the written word permitted him, as he put it, 'to ensure a dignified beat of the mandarin's fan'. By such means he was able to put into wider currency his contempt for literary figures he considered to have been overrated: and the world was rebuked for its supposition that Dostoevsky, Stendhal, Conrad, D.H. Lawrence, Gide, Sartre, Thomas Mann, Pound, Eliot, or Faulkner were worth bothering with; it was advised to stop reading *Dr Zhivago* and to start reading H.G. Wells's *The Passionate Friends*.[1] Some resented the dismissiveness of his magisterial pronouncements, others seemed keen to encourage such stage-managed 'outbursts'. However unjust some of his strong opinions may strike us as having been, many will sympathise with one or two of his nominees for the hall of fame unmerited; and there is an undoubted value in fluttering academic dovecots by challenging the ranking system: if nothing else, Nabokov's scorn can send us to discover whether a writer really is as awful as he claims.

1. His biographer does not endorse Nabokov's judgement of Wells's book; see Boyd, p. 91.

This boisterous scepticism must of course also be applied to Nabokov's own writing. *Speak, Memory* is highly regarded by nearly everybody, but with the exception of *Lolita*, on whose excellence there is something like general agreement, there is an interesting and possibly symptomatic lack of consensus about what constitutes his best work in fiction. Which novels represent the true Nabokovian canon? *The Gift* is usually cited as his major novel in Russian, but even here there are some problematical aspects.[2] That aside, in the corpus of his fiction in English the case of *Ada*, which he seems quite consciously to have devised as his *magnum opus* (it is comfortably his longest novel), most clearly illustrates divergent judgements: by some commentators the book is accepted as the *chef d'oeuvre* it impersonates, whereas for others it appears to be a colossal misjudgement on the part of its author. After the engorgement of *Ada*, *Transparent Things* offered a curiously slight fictional edifice, which was so unappreciated by its readers that Nabokov was driven to explain to them what the book had been about ('a beyond-the-cypress enquiry into a tangle of random destinies'); *Look at the Harlequins!*, which was a diminished mirror-version of Nabokov's own life and achievements, displayed one or two familiar tints, but on the whole was greeted as a disappointing performance. It may well be that Nabokov's Swiss eminence was less fertile for his gifts than had been his American obscurity: America is marginal to these last three novels, and nothing in them offers an equivalent resistance to the self-gratifications of imagination – *Ada* takes place in a realm called 'anti-terra', an inversion of the 'real' world that is in its most interesting aspects a kind of hallucinated image of pre-Soviet Russia. We miss John Shade's sobriety to counterbalance this excess; and the notes which Nabokov appended to paperback editions of *Ada* resemble self-advertising cleverness rather than the mock-pedantry of Kinbote.

They may also imply an authorial reluctance to grant his readership any rights of misunderstanding. 'I work hard, I work long, on a

2. The problems may be of my own making, but I am unable to read *The Gift* and feel either that its hero Fyodor is touched by genius, or that he had a loving relationship with his father, or that his father is a noble human being, or that Fyodor is in love with or is loved by Zina Mertz (whom he first hears using the lavatory in the boarding house they live in). All these aspects seem to me to be the object of an authorial corrosion on Nabokov's part, producing in Fyodor a parodically diminished version of his own life. Only Long, amongst the critics with whose work I am familiar, sees the novel in a similar way to me; so perhaps it is our *folie à deux*.

body of words until it grants me complete possession and pleasure', he declared; and taken together with his assertion that an artist imagines his audience – if at all – as a roomful of people wearing his own mask, we may wonder whether his desire to dominate extended beyond his fiction to its anticipated readers. It is probably the case that (theories of 'the death of the author' notwithstanding) a reader stands in relation to a fiction-writer as Ariel does to Prospero: an enabling but subordinate spirit. It can at times appear that a reader inhabits Nabokov's enchanted island on sufferance, a necessary Caliban whose only function is to do his master's bidding and succumb to his superior magic; for the creative latitude exemplified in Kinbote's strong misreadings is seldom granted to Nabokov's readers, for whom *Pale Fire*, architectonically perfect, is a text permeated by the consciousness of authorial preordination, in which nothing can be discovered which has not been master-minded, and even the digressions have been calculated. Seeking to follow the advice of the eccentric great-aunt in *Look at the Harlequins!* ('Play! Invent the world!'), Nabokov's reader is apt to find that wily Prospero has already invented it for him, and that the rules of the game resemble Humbert's predatory chess as much as Lolita's disinterested tennis. The image of those interviews whose spontaneity was scripted comes to mind, as a faintly disturbing model; one thinks of Milan Kundera's definition of 'graphomania' as a mania to write books, 'the mania not to create a form but to impose one's self on others. The most grotesque version of the will to power'.[3] There are certainly aspects of a will to power in Nabokov's writing – not least in his edition of *Eugene Onegin*, in which he set out to be the permanently dominant authority on Pushkin, a scholar who could not thereafter be ignored.

The charge that has sometimes been levelled at his fiction is of a certain human attenuation, a lack of copiousness or incidental detail particularly evident in the late work. For some readers *Pale Fire* seems more like the working model of a novel than the thing itself, a triumph of technique rather than a visceral experience. Nabokov professed amazement at authors whose characters developed in ways unforeseen by them, and described his own position as creator in terms of an absolute tyranny; yet for all his scorn for a writer such as Faulkner, he did not create a character so vividly imaginable, so suffused with the illusion of life, as Jason Compson in *The Sound and the Fury,* or Anse Bundren in *As I Lay Dying.* In

---

3. See 'Sixty-three Words', in Milan Kundera, *The Art of the Novel,* tr. Linda Asher, (London: Faber and Faber, 1988), p. 131.

spite of accurate attention to the material world, and his scientist's familiarity with its flora and fauna, Nabokov lacked the creative immoderation of an author like Dickens, who could hardly bear to rein in his imagination, even at the risk of going too far with a description; for his affinity was rather with careful Ben Jonson, than with the Shakespeare whom Jonson criticized for not blotting out more lines. Yet in Nabokov's literary criticism the 'human interest' he so deplored keeps breaking in to guide his responses: as when he addresses the 'child theme' in *Bleak House,* or describes the selfish use made of Gregor Samsa by his family in Kafka's story, or indignantly denounces the persecutors of Don Quixote. Of course, Nabokov was also unrelenting in his attack on those for whom art is a matter of 'sincerity' and 'simplicity' rather than of deception and calculation; we might well remember Pope's couplet: 'True ease in writing comes from art not chance, / As those move easiest who have learned to dance'. Nabokov believed that 'art at its greatest is fantastically deceitful and complex'; and if at one level the deceiver exerts a dishonest power over his victim, in the case of reader and author the former has not had a fiction imposed on him, but has collaborated with its illusion, and knowingly accepts the fiction for what it is (remembering Wallace Stevens's dictum, with which Chapter Two closed). Rather than hypnotist and subject, then, the truer model for the relation between Nabokov and his reader is surely that of a civilised *rapport,* in which one joins him on a fictional plateau, on a high terrace of consciousness, for the sake of its view. Just as his lectures showed him to believe the literary classics to be common property, a shared experience which is made meaningful to the degree that the reader individualises his response, so as a writer Nabokov paid his readers the compliment of supposing they could almost be his equals; collaborators rather than antagonists.

It is a complex ascent to that plateau, however; and some commentators have been apt to preen themselves too much on having managed it, seeming anxious to proclaim their solutions to Nabokov's puzzles as if in itself that were a sufficient critical response. If his playfulness was sometimes tinged with pedantry, so was his pedantry lit by playfulness (as in the note in *Eugene Onegin* describing a Professor Snegiryov as 'ethnographer, censor, and fogy'). In spite of the satire of sycophancy offered in the Kinbote/Shade relationship, Nabokov has attracted some Kinbotesque attentions and some perhaps-too-compliant commentators. But these do not lie near the heart of what is valuable in his work, which is the celebration of consciousness enfranchised in fiction; such a mood as Henry James expressed in an eloquent memorandum (more a declaration of faith) in his Notebooks:

> To live in the world of creation – to get into it and stay in it – to frequent it and haunt it – to *think* intensely and fruitfully – to woo combinations and inspirations into being by a depth and continuity of attention and meditation – this is the only thing.

Little as he liked James's work, Nabokov would surely have acknowledged some fellow-feeling with this description of a specialized absorption in the creative occasion: an absorption experienced by the reader as well as the writer. This naturally led to Nabokov's distrust of generalized discussions of literary periods, and even of an author's *oeuvre*; he preferred instead to focus on the single work, as the force-field of imagination in which this otherworld was generated. In his commentary to *Eugene Onegin* he inveighed against the various '-isms' of literary history, its talk of 'schools' and 'movements', which he condemned as 'harmful chiefly because they distract the student from direct contact with, and direct delight in, the quiddity of individual artistic achievement (which, after all, alone matters and alone survives)'.

Any assessment of Nabokov's artistic achievement must start with his undoubted brilliance as a prose stylist; the question is whether it also ends there. The formal effectiveness of his novels and their games-playing has been acknowledged from early on; but in spite of his detestation of writing which has a message to deliver or a tub to thump, more recent criticism has attempted to locate a continuous human concern and even a justifying metaphysics underlying the immaculate performance. Nabokov himself prefigured such a line when he published his belief that 'one day a reappraiser will come and declare that, far from having been a frivolous firebird, I was a rigid moralist kicking sin, cuffing stupidity, ridiculing the vulgar and cruel – and assigning sovereign power to tenderness, talent, and pride' (SO p. 193). The proud man ridiculing the vulgarian is a rather small-scale moral hero, we might think; a step away from equating bad taste with bad faith. When Nabokov addressed himself directly to ethical matters his vocabulary could be curious: his considered reply to the question 'what is the worst thing men do?' was, 'to stink, to cheat, to torture'. Readers who imagined that his hostility to dictatorships would necessarily extend, in an American context, to a committedly 'liberal' politics, did not know how undisconcerted Nabokov had been by the McCarthy investigations into 'unAmerican activities': when the blurb-writer for Penguin's edition of *Pnin* defined McCarthyism as a 'Nabokovian enemy' the mistake was glaring, since in that novel the professor who alludes sarcastically to the witch-hunting senator is himself satirically undermined. In *Pale Fire* 'pink' politics are mocked; in 1965 Nabokov, who normally abstained from such displays, sent President Johnson a get-well telegram endorsing his policies (which included escalation of the Vietnam War as well as desegregation). One is tempted to suggest that nonfictive things stink most when writers wink.

D. H. Lawrence famously advised readers to trust the tale rather than the teller: can we deduce an ethics from the aesthetics of Nabokov's highly-wrought fictions? Kundera has written of the inherent opposition between the spirit of the novel with its 'radical autonomy', and the 'Totalitarian Truth', to which the novel's negative capability is anathema.[4] As the playground of extended consciousness operating with no practical end in view, the novel necessarily gainsays systems of use, and celebrates the pure excessiveness of being human – it is moral, in Nabokov's view, to the degree that it is uninhibited by considerations from outside the realm of fiction. He was also deeply concerned with constructive memory, and meditated on the absurdity of a life in which what has happened simply disappears, as if of no further relevance: his fiction constantly evokes the chasm between past and present, to deny it – to refuse to believe all our yesterdays lead nowhere, that the dead are forgotten, that our memory of what has been and our imagination of what may be are sham. As he acknowledged, his was a protest against time – although the more he discussed it directly as a subject (as in parts of *Ada*), rather than evoking its interwoven textures or disrupting its linearity in his fiction, the less effective his protest tended to be.

In his recent biography, Brian Boyd has suggested that Nabokov's concern with consciousness and what it portends is a fundamental of his fiction, a metaphysical preoccupation that implies a kind of private theology: a sense of behind-ness or beyond-ness that has relevance to the false fronts of many of his novels. We cannot, however, with safety allow their author's philosophical position to undermine our sense of the novels' quiddity. Doubtless one of the reasons he admired Pushkin's masterpiece was that, being a verse-novel with a fairly complex stanza, it constantly foregrounds its literariness, and a reader cannot sensibly discriminate its matter from its manner. In doing this, Pushkin displays simultaneously the conjuring trick (the story of Onegin and Tatiana) and the means by which it is performed (the obtrusive form), without diminishing the magical effectiveness.

---

4. In his essay 'The Depreciated Legacy of Cervantes', in *The Art of the Novel* (see previous note).

With their calculated artistry Nabokov's novels as consistently enforce their difference from unstructured 'life', and if we consciously accede to their spell-binding may reveal to us aspects of our humanity, as we explore through them the pleasures and compulsions of our appetite for fiction.

The fly-leaf of Boyd's substantial first volume announces Nabokov as 'one of the greatest writers of this century'. There will probably be many dissenters from this view; in any case we are too close to judge, and History will decide the question in her own time. For my part, I suspect that he will find his place below Jane Austen, whom he slightly patronized, but above Robert Louis Stevenson, whom he slightly overvalued. Whatever the eventual state of his reputation as a writer, his contribution to the art of fiction can hardly be ignored, and should not be undervalued: there are some bright jewels in his treasury, and the muses must have gathered expectantly around the waiting cradle when, on 23 April (New Style) 1899, as dawn broke over St Petersburg, Vladimir Nabokov was born.

Chronological Table

Note: *Nabokov's publication history is complex. In the following table, dates are given for the first publication in book form of his novels, which is not in all cases a reliable indication of the date of composition: those wishing to establish the detailed sequence should consult Boyd. Dates of translations of Russian works are given in square brackets, together with any subsequent revisions; details of translators etc. may be found in Grayson (see Bibliography). Nabokov's later work was published in the US before the UK.*

1899	23 April (New Style): Born in St Petersburg.
1916	Privately publishes his first volume of poems.
1919	15 April (New Style): Leaves Russia, never to return.
1919–22	At Trinity College, Cambridge.
1920	August: V.D. Nabokov (father) moves his family from London to Berlin.
1922	28 March: Assassination of V.D. Nabokov in Berlin.
1922	June: Graduates from Cambridge with Second Class Honours in French and Russian.
1922	Summer: Commissioned to translate *Alice in Wonderland* into Russian.
1923	December: Elena Nabokov (mother) moves to Prague, with her younger children. VN remains in Berlin.
1925	15 April: Marries Véra Slonim, in Berlin.
1926	First novel published in Berlin, *Mashen'ka* [*Mary*, 1970].
1928	*Korol' Dama Valet* [*King, Queen, Knave*, 1968].
1930	*Zaschita Luzhina* [*The Defence*, 1964].

1932	*Podvig* [*Glory*, 1972].
	Kamera Obskura [Camera Obscura, London, 1936; *Laughter in the Dark*, New York, 1938, rev. 1961].
1934	10 May: Birth in Berlin of only child, Dmitri.
1936	*Otchayanie* [*Despair*, London 1937; New York, 1966]
1937	18 January: Leaves Berlin, and Germany, forever.
1937	11 February: At short notice gives a reading of his work in Paris, replacing an indisposed Hungarian writer; James Joyce is in the audience.
1937	February: In Paris, commences a love affair with Irina Guadanini, which VN terminates in September.
1937	Late April: Wife and son leave Berlin for Prague, where VN joins them on May 22. At the end of the following month the three of them move to France.
1938	*Sogliadatay* [*The Eye*, 1965].
	Priglashenie Na Kazn' [*Invitation to a Beheading*, 1959].
1938	Autumn: Settles with his family in Paris.
1939	February: Is invited by mutual friends, the Léons, to dine with Joyce; the evening is unmemorable.
1940	May: Leaves France for the USA, together with his wife and son.
1940	Meets Edmund Wilson, through whom he makes some useful contacts, and with whom he enjoys a real – if adversarial and finally acrimonious – friendship.
1941	*The Real Life of Sebastian Knight*.
1941	Summer: Lecturer in Creative Writing, Stanford University.
1941–48	Part-time Research Fellow in Entomology at the Museum of Comparative Zoology, Harvard University.
Fall 1941 – Spring 1942	Visiting Lecturer in Comparative Literature, Wellesley College.
1943–48	Temporary Lecturer in Russian Language and Literature, Wellesley College.
1944	Publishes his study of Nikolai Gogol.
1947	*Bend Sinister*.
1948–59	At Cornell University, eventually as Professor of Russian and European Literature.
1951	*Conclusive Evidence* (later revised as *Speak, Memory*) Spring semester, 1951–2 Given leave of absence

	from Cornell to deliver his lectures on *Don Quixote* at Harvard.
1952	*Dar* [*The Gift*, 1963]; the first full publication of Nabokov's last novel in Russian, which when published serially in the émigré journal *Sovremennye Zapiski* (1937–8) omitted the fourth chapter (Fyodor's biography of Chernyshevski), at the editors' insistence.
1955	*Lolita* (Paris).
1957	*Pnin.*
1958	*Lolita* (New York).
1959	Resigns professorship at Cornell; travels in Europe, returning to USA for an extended visit in 1960, spent mostly in California working on a screenplay for the film of *Lolita*.
1961	Settles at the Palace Hotel, Montreux.
1962	*Pale Fire.*
1964	Publishes his translation and commentary on *Eugene Onegin*. Edmund Wilson's hostile review initiates the public deterioration of their friendship.
1966	*Speak, Memory.*
1969	*Ada.*
1972	*Transparent Things.*
1974	*Look at the Harlequins!*
1977	Dies in hospital at Lausanne, Switzerland.

Bibliography

Writing by Nabokov

The enthusiast will want of course to read all of Nabokov's novels, short stories and (perhaps) poetry, which are now available in translation where not originally composed in English. Of the novels in English not covered by this study, *Ada* is probably the one to read next. I would advise the reader wishing to explore the Russian novels to leave *The Gift* until he has read some of the earlier work – for example *Mary* and *The Defence* – as well as Nabokov's literary criticism on Russian writers.

Although later he referred to it disparagingly, Nabokov's study *Nikolai Gogol* (New Directions 1944, 'corrected edition' 1961) is entertaining about its subject, and a useful introduction to Nabokov's work as well. His *Lectures on Russian Literature* (Picador 1983) repeats some of the Gogol material, and adds to it discussions of Turgenev, Tolstoy, and Chekhov (among others) which again throw light on their writing and Nabokov's. His *Lectures on Literature* (Picador 1981) addresses fiction by Jane Austen, Dickens, Flaubert, R. L. Stevenson, Proust, Kafka and Joyce; this volume interestingly illustrates the approach Nabokov took to literary study. The companion volume, his *Lectures on Don Quixote* (Weidenfeld and Nicolson, 1983) is a good deal less illuminating, but worth seeking out.

Speak, Memory is an essential text. *Strong Opinions* (Weidenfeld and Nicolson 1974) also contains insights into Nabokov's views on fiction, his own and others', and glimpses of his Swiss existence. *The Nabokov-Wilson Letters* (Weidenfeld and Nicolson 1979) has much to reveal about Nabokov's American years, as well as illustrating the kinds of disagreement about literary values and Russian history to which he often found himself a party; in addition, it contains a very useful introductory essay by the editor, Simon Karlinsky, outlining the nature of St Petersburg at the turn of the century. *Vladimir Nabokov: Selected Letters 1940–1977* (Weidenfeld and Nicolson 1990), co-edited by his son Dmitri, adds further detail to the portrait of this period.

Literary Contexts

These of course are wide-ranging; but anybody wishing to explore Nabokov needs to have some notion of the literature he himself valued, particularly in Russian. Pushkin's *Eugene Onegin* comes top of the list; what is generally acknowledged to be a fine translation into English, which reproduces Pushkin's stanza without gross distortions of sense or accuracy, has been made by Charles Johnston (Penguin, 1979). Those with access to a university library or its equivalent will want to read this side by side with Nabokov's own literalist four-volume translation and commentary (1964; revised edition Routledge and Kegan Paul 1975). Gogol's *Dead Souls*, Tolstoy's *Anna Karenina*, the short stories of Chekhov, and Bely's *Petersburg* complete this (barbarously abridged) list of Russian works Nabokov admired. All are available in Penguin, although not necessarily in translations which Nabokov would have approved.

Outside Russian literature, Sterne's *Tristram Shandy*, Flaubert's *Madame Bovary*, and Lewis Carroll's 'Alice' stories are relevant. Kafka's *Metamorphosis* and Joyce's *Ulysses* were greatly admired by Nabokov, at the same time that he disclaimed any influence. Indeed, none of the above works should be taken directly to have influenced Nabokov; but the reader will probably understand more about him by reading them than by reading monographs about Nabokov.

Criticism on Nabokov

In common with most significant writers and some insignificant ones, Nabokov has spawned a critical industry, embracing a specialist newsletter and Nabokov conferences (although it is difficult to imagine a more inherently un-Nabokovian activity than this). There is much more written on Nabokov than is worth reading (with Kinbotian modesty I leave my reader to pass appropriate judgement on *this* book); therefore I shall be rigorously selective in what follows.

Having chosen his own biographer against advice, Nabokov's last years were spent disputing and attempting to correct the resulting book, *Nabokov: His Life in Part* by Andrew Field. If all this had a smack of Kinbote and John Shade, the appearance of a subsequent biographer, and the interchanges between him and Field, at times seemed reminiscent of the struggle between Esau and Jacob. It is generally agreed that Brian Boyd now holds the field; the first volume of his two-volume biography *Vladimir Nabokov: The Russian Years* (Chatto and Windus 1990) is an indispensable factual record, which contains as well acute commentary on the fiction of the period dealt with. His second volume will directly address the story of Nabokov's relations with Field; meanwhile the curious reader may wish to cross-check with Field's *VN: The Life and Art of Vladimir Nabokov* (Futura 1988), to get the flavour of it all.

Jane Grayson's *Nabokov Translated: A Comparison of Nabokov's Russian and English Prose* (Oxford University Press 1977), is very useful for those who, like me, have no Russian and wish to know the extent to which his Russian novels were revised in translation. Alfred Appel's *The Annotated Lolita* (Weidenfeld and Nicolson 1971) solves a lot of textual problems. Norman Page's *Nabokov: The Critical Heritage* (Routledge and Kegan Paul 1982) shows how the works have been received on publication.

There are some useful items amongst the two following collections: *Nabokov: Criticism, reminiscences, translations and tributes,* edited by Appel and Newman (Weidenfeld and Nicolson, 1971), and *Vladimir Nabokov: A Tribute,* edited by Peter Quennell (Weidenfeld and Nicolson 1979). Some of the earlier studies of Nabokov were perhaps too deferential and over-concerned with the formal aspects of his fiction; in various ways the following books rectify those imbalances: G.M. Hyde, *Vladimir Nabokov* (Marion Boyars 1977); Ellen Pifer, *Nabokov and the Novel* (Harvard University Press 1980); David Rampton, *Vladimir Nabokov: A critical study of the novels* (Cambridge University Press 1984); Michael Long, *Marvell, Nabokov: Childhood and Arcadia* (Oxford University Press 1984; I should perhaps note that I do not select this merely because the author refers to me in his Acknowledgements). I am not in complete agreement with any of these studies, of course; but together with Boyd's biography they have seemed to me the most stimulating criticism of Nabokov. There are good bibliographies in both Rampton and Long, and there is promised to be one in Boyd's second volume.

Index

NB *Authors' pseudonyms are used, where applicable. Nabokov's novels are indexed under their English titles only.*